Mesolithic Exploitation Patterns in the Central Pennines

A Palynological Study of Soyland Moor

Christine T. Williams

with a contribution by
V. R. Switsur

BAR British Series 139
1985

B.A.R.

5, Centremead, Osney Mead, Oxford OX2 0ES, England.

B.A.R. 139, 1985: 'Mesolithic Exploitation Patterns in the Central Pennines'

ISBN 9780860543299 paperback
ISBN 9781407317519 e-book

DOI https://doi.org/10.30861/9780860543299

A catalogue record for this book is available from the British Library

This book is available at www.barpublishing.com

CONTENTS

LIST OF ILLUSTRATIONS AND TABLES

ACKNOWLEDGEMENTS

I am indebted to the Department of Education and Science for providing the grant to enable me to undertake this research, to Professor A J Willis for making facilities available in the Department of Botany at Sheffield University, and to my supervisors Dr P A Mellars and Dr B D Wheeler for their long-suffering support and advice. Dr V R Switsur of the University of Cambridge generously provided the eleven radiocarbon dates which greatly helped in the interpretation of my data.

A prehistorian by training, I am indebted to many botanists and geographers who helped me to understand the techniques and principles of palynology and ecology. The pollen site of Soyland Moor was pointed out to me after much fruitless searching by Dr D D Bartley of the University of Leeds. Dr Bartley also helped me to devise and undertake my sampling strategy and identified the peat composition. My grateful thanks are extended to Dr P D Moore of King's College, London, without whom this research would not have been completed. On my numerous visits to London, Dr Moore taught me to identify pollen, determined the more obscure grains, gave me the run of his laboratories and reference collection, and gave me much advice on practical and interpretative problems. A special thanks is also extended to Dr J H Tallis of the University of Manchester who generously made available an unpublished diagram from his site of Robinson's Moss in Longdendale which allowed me to corroborate much of the Soyland Moor data. Dr Tallis also determined the amount of charcoal present on some of my pollen slides, and discussed many aspects of my diagrams with me.

I would also like to thank Miss K Hattersley for providing the section through the mire, and Miss J C Bird for her speedy and efficient typing.

SUMMARY

Palynological investigations from the mire on Soyland Moor, on the central Pennine plateau, near Ripponden, were undertaken to study the economic strategies of Mesolithic peoples. A small mire began to form at Soyland in the early postglacial (c 7000 bc), and the mire is situated at the northern edge of the most densely occupied area of Mesolithic Britain, thus optimising the possibilities of producing the type of data required.

The resulting pollen spectra were firstly interpreted as representing the natural postglacial vegetation development, with possible regional variations discussed in comparison to other sites. The use of four core sites from one mire allowed some tentative interpretations of the geographical distribution of the vegetation to be made. Analysis of the data in anthropogenic terms showed that man may have extensively manipulated his environment during the Late Mesolithic, but in a different and more intensive manner than that postulated by current hypotheses. Two long periods of clearance activity (over 500 years) were perceived, where artificially heightened grazing pressure produced a 'parkland' type of vegetation. Pre-Neolithic arable cultivation at the very early date of 3870 ± 95 bc was also indicated. Possible long-term effects of the alleged clearances were seen in the early development of heathland, and the extension of the mire in late Zone VIIa. The regional variations of the postglacial woodland were re-interpreted as the result of anthropogenic activity, as were the Elm Declines. The final discussion relates the palynological data to the archaeology, particularly on their relevance to culture change processes. In the light of the Soyland Moor evidence, it is difficult to sustain the hypothesis of invasion, still currently in vogue as the explanation of the spread of agriculture in Britain.

Abbreviations and Conventions

SMA	Soyland Moor site A
SMB	Soyland Moor site B
SMC	Soyland Moor site C
SMD	Soyland Moor site D
TP	Total pollen
AP	Aboreal pollen
NAP	Non-arboreal pollen, divided into dwarf shrub and herbaceous pollen
b.c/b.p	Uncalibrated radiocarbon years

CHAPTER 1

INTRODUCTION

The data presented here constituted research undertaken for a postgraduate degree at the University of Sheffield, arising out of an interest in Mesolithic economy, and specifically the way in which these peoples manipulated their environment to provide predictable and adequate food supplies. The hypothesis of Mesolithic clearance has become an accepted part of the archaeological lexicon, and the most fruitful area for finding evidence to support the hypothesis has been palynology. However, while accepting both the principle and the evidence, it was felt that the hypothesis had by no means been fully explored nor the available data been studied in a detailed and critical manner. The search for clearance evidence has been haphazard, and few new sampling strategies have been devised to specifically test the hypothesis. The mechanisms of clearance have not been explained in any great detail, either archaeologically or botanically. Lastly, the plethora of ethological and modern botanical data used to buttress the hypothesis has been obtained from environments significantly different to those of postglacial Britain. These data may merely serve to confuse the issue, and can certainly mask a lack of empirical evidence. This research aimed to devise sampling strategies to optimise the appearance of clearance evidence in pollen diagrams. If this evidence was forthcoming I hoped to be able to interpret it in management terms: the aims and mechanisms of the clearance, and its scale in time and space were to be studied. The possibilities of gaining such evidence from raw pollen data were to be explored. Finally, a study of how such economic strategies would have affected Mesolithic society and their relevance to the causes and mechanics of culture change was undertaken.

The central Pennine area was chosen for a palynological study. This area is defined here as the area above 300 m, loosely bounded to the north by the Aire gap and to the south by Longdendale, following Barnes (1982). Some authors refer to this area as part of the southern Pennines, along with the whole of the Peak District (eg Tallis 1964a and b). Here, the term central Pennines is preferred, as it has a different archaeological and ecological history to the Peak. This area was chosen for study because it is exceptionally rich in Mesolithic sites, and may well represent the most densely occupied area in Britain during this period (Mellars 1974). This area is regarded as worthy of further study, especially as most past research has been left to amateur archaeologists. The main problem in studying these Mesolithic sites is the paucity of economic evidence, only flints and a few hazelnut shells being recovered (Clark 1932). Pollen analysis has helped to fill this gap. Many pollen analyses have already been undertaken in the area (eg Tallis 1964a, 1964b, and Bartley 1964, 1975), but most of them do not cover the early post-glacial period correlating with the Mesolithic. Also, the aims of the palynologist are not always consonant with those of the archaeologist. Even where pollen analysis has been undertaken on archaeological sites, the results have sometimes been tucked away in appendices and not fully integrated into the archaeological discussion of economy and culture change (eg L Brown in Barnes 1982). This research was undertaken to provide palynological information with direct application to the archaeology.

The first task was to reconstruct the local vegetation sequence of the early postglacial period, using the Elm Decline as a convenient, if rough, correlation with the end of the Mesolithic. This would provide a basic framework in which to place the archaeological record and to study regional differences with other areas of Britain. Once accomplished, evidence for man's manipulation of the environment could be looked for and analysed, as outlined above. If this evidence was found, a number of further issues worthy of discussion would arise. The way in which such activity affected the vegetation composition in the long term was of great interest, particularly as vegetation succession can cause or hasten culture change. The extent of the woodland and heathland cover through time was of particular importance here. Despite the current acceptance of the Mesolithic clearance activity in the literature, many researchers feel that the supposed small population of this period could not have caused any large scale or permanent changes in the vegetation (eg Simmons 1981). This cannot be accepted on a priori grounds, as population size is dependent on the intensity and success of its economy (Higgs and Jarman 1975). Besides, even small populations can have disproportionate effects on their environment (eg Turner 1970), particularly in marginal areas near to the threshold of change. Jacobi et al (1976) have shown that man may indeed have had a great effect on the pre-Neolithic landscape of the central Pennines, particularly in the extent of woodland cover. Peat initiation, too, may have been a result of, or precipitated by, man's activities (Tallis 1975, Moore 1975a). The Elm Decline was considered to be as much a product of the Mesolithic as of the Neolithic and thus needed consideration here. If the Elm Decline was a by-product of Neolithic agriculture practices, as many researchers believe (eg Garbett 1981), it has to be remembered that this type of economy did not spontaneously generate in a vacuum, but was either evolved from or integrated with the Mesolithic economy and society. The Elm Decline has to be seen in the light of five millennia of Mesolithic occupation.

In order to produce data which could help to elucidate these problems, pollen deposits had to be found which satisfied a number of requirements. These were:

1. Mire deposits were preferable to lake deposits, as there is little problem of pollen inwash (Pears 1977), and less possibility of mixing of deposits (Beckett 1979).

2. Early peat formation was necessary, preferably from the late glacial period, in order to cover the whole of the Mesolithic.

3. Fast peat growth throughout the period in question was desirable to allow good sampling resolution and to reduce the possibility of an hiatus in peat accumulation.

4. A small mire was desirable, so that the pollen rain was mainly local (Janssen 1966). This helps to optimise the appearance of clearance data (Smith 1982).

5. The mire had to be close to areas of Mesolithic habitation, again to optimise the chances of clearance evidence appearing in the diagrams. In the central Pennines, this is not very difficult as long as the mire is situated between 350-500 m, which is where the

majority of sites are located.

These requirements were difficult to fulfil and six months of futile searching followed. Finally, Dr D D Bartley of the University of Leeds suggested a site on Soyland Moor which was known to have peat over 7 m deep and therefore may have been able to fulfil the requirements. Gross sampling proved that, indeed, this mire started to form early in the postglacial, probably as far back as Godwin's Zone IV, and also that it had been less than 300 m in diameter for much of the pre-Elm Decline period. At 389 m, it was surrounded by Mesolithic sites. Thus, Soyland Moor seemed to provide most of the features necessary to maximise the chances of finding the information required.

Soyland Moor is situated on the high millstone grit plateau watershed of the central Pennines. It is situated north of the A58 Halifax to Bolton road and straddles the Greater Manchester/West Yorkshire boundary (National Grid Reference SD 9718). Rainfall is high at c 1542 mm per annum (Crump and Crossland 1904). The geology of the area is relatively simple: the central Pennines are composed of millstone grit interspersed with soft shales, bordered to the north and south by the carboniferous limestone of the Yorkshire Dales and the White Peak, and flanked by the lower coal measures (Wray *et al* 1930). The watershed is a broad plateau usually above 330 m, and areas above 330 m are referred to as 'upland' throughout. This plateau has a 'stepped' appearance caused by the differential erosion of the hard grits and soft shales (Barnes 1982). The gritstone plateau is deeply dissected by narrow steep-sided valleys known as 'cloughs' (Smith and Moss 1903), which provide shelter for tree growth today. The other features of this landscape are small, well-drained west-facing escarpments (Barnes 1982). South of the Aire-Calder watershed and above 300 m there is no evidence for glaciation (Wray *et al* 1930, Raistrick 1934). If not an actual nunatak, the central Pennines constituted a peninsular jutting north into the surrounding ice during the maximum extent of last glaciation.

Today, the area is floristically poor due to the base-poor parent material, and perhaps to management practices and pollution (Pearsall 1968). Blanket bog covers much of the area above 350 m and is today dominated by *Eriophorum vaginatum* and *Molinia caerulea*, as, indeed, it has been for this century at least (eg Crump and Crossland 1904). Peat stratigraphy does indicate, however, that at periods in the past *Sphagnum* was the dominant bog-builder (eg Tallis 1964c). Woodland rarely occurs on the exposed plateau top, but can reach up to 330 m in the shelter of the cloughs (Woodhead 1906, Crump and Crossland 1904). This woodland is composed mainly of oak and birch, interspersed with rowan, willow, holly and blackthorn (Smith and Moss 1903).

Soyland Moor itself is composed of upper Kinderscout grit. To the immediate south, Blackstone Edge rises to 472 m OD, and to the north-west Byron Edge rises up to 400 m. Beyond Byron Edge, the land sharply drops into Turvin Clough which bounds the west and north of Soyland Moor. To the east, the Moor slopes gently down to the head of the steep valley of the major drainage channel of the area, the River Ryburn. All of Soyland is peat-covered today, but only a small area is characterised by deep peat. Blackstone Edge Reservoir may well have been part of this area (Mr R Wood, Yorkshire Water Board, pers comm). Otherwise, the peat is rarely deeper than 2 m, and is dissected into hags by small drainage channels on the sloping part of the Moor to the east. *Eriophorum angustifolium* is the dominant species over the Moor, with some *E. vaginatum* and *Molinia caerulea*. To the south, Blackstone Edge is too steep to support more than small pockets of peat, and is dominated by *Vaccinium myrtillus*. The area of deep peat on Soyland appears to have accumulated in a shallow basin at the confluence of two streams, representing either a small open-ended lake, or a constantly waterlogged flood plain (D D Bartley, R Smith and P D Moore, pers comm). This topogenous mire began to accumulate in the early postglacial period and finally merged with the blanket bog which started to form in the immediate pre-Elm Decline period across the plateau. Such early peat formation is unusual in the central Pennines, and only Robinson's Moss in Longdendale, 20 km to the south (Tallis unpublished) has a similar early date. These two sites are on the north and south boundaries of the area where most Mesolithic sites are to be found (Barnes 1982). North of Soyland, sites are plentiful but not as abundant as to the south.

The above gives the background to the site involved in the research. The following gives a brief resume of the prehistory of the Mesolithic in Britain, and the central Pennines in particular, in order to provide the archaeological framework in which the palynological research took place.

CHAPTER 2

THE ARCHAEOLOGY OF THE MESOLITHIC

The Mesolithic period has often suffered in comparison to the Upper Palaeolithic and the Neolithic. The Upper Palaeolithic is regarded as a brilliant culmination of man as hunter, with his highly specialised techniques of adapting to the hostile environment of the late glacial, the period when he attained a high cultural status as shown in cave and mobiliary art. The Neolithic saw the widespread development of agriculture, a settled way of life and the construction of megalithic monuments. This has been regarded as a cultural progression from hunter-gathering (eg Childe 1947). By contrast, the Mesolithic produced very little in the way of remains: sites were small and ephemeral (although more abundant than from any other period), and consisted of very little other than tiny worked flints and waste. In comparison to the Upper Palaeolithic caves, with their wealth of bones and carefully worked tools, and to the Neolithic sites, with their readily recognisable evidence for agricultural and ritual practises, not only did the Mesolithic evidence appear meagre, but it provided less scope for study, and therefore for understanding.

The Mesolithic period has also suffered a historical hangover. Until early this century, it was thought that an hiatus existed between the two pillars of the Palaeolithic and Neolithic (Clark 1978). Thus, even when the overwhelming accumulation of evidence finally convinced archaeologists that the Mesolithic was a reality, its status as a stop-gap remained, a cultural degeneration from the Upper Palaeolithic and a waiting period before the dynamics of agriculture began (Clark 1978).

The recognition and documentation of the importance of the Mesolithic in Britain was accomplished almost single-handedly by J G D Clark, who produced a mass of evidence on the flint assemblages (eg Clark 1932, 1936) and the economy (eg Clark 1952, 1954). However, despite our greatly increased knowledge of the period, the old attitudes still prevailed in certain quarters. Thus, the Mesolithic was still "a stage of 'waiting' or 'quarantine' which was realised in an environment where primitive agriculture could not develop." (Kozlowski 1973: 333). This greatly underestimated the successful adaptation of Mesolithic man for over five millennia (half the length of the postglacial period to date) in the face of rapidly changing environmental conditions. Ethnography has shown us that an impoverished material culture does not equate with an undeveloped society (*Man the Hunter* by Lee and DeVore (1968) being the classic example). However, it is unprofitable to separate the Mesolithic entirely from its preceding and succeeding periods, except as a useful typological label. While continuity of peoples from the Upper Palaeolithic through to the Neolithic is difficult to document (Morrison 1980), dislocation of settlement patterns and changes in tool kits are, on *a priori* grounds, more likely to represent adaptations to changing circumstances than to the influx of new peoples (Binford 1968, Clarke 1973). Thus, it may be that the same people who made the sophisticated Upper Palaeolithic tool kits and herded reindeer (Sturdy 1975) also successfully coped with the rapid environmental changes of the postglacial and finally became farmers.

The concepts of degeneration and progression are irrelevant to the process of change.

The Mesolithic in Britain

By convention the start of the Mesolithic period coincides with the early postglacial. Most late Upper Palaeolithic sites are dated to pre-10,000 bc (Mellars 1974), but some later survival is seen. For example, Robin Hood's Cave at Creswell Crags in Derbyshire has a late Upper Palaeolithic assemblage associated with a date of 8440 ± 90 bc, and Anston Cave in South Yorkshire has dates as late as 7800 ±110 bc (although it has been argued that these dates may be too young (Mellars 1974)). The earliest dated Mesolithic site is Thatcham site III in Berkshire at 8415 ± 170 bc (Godwin and Willis 1964). This possible overlap may indicate continuity of occupation, with the Upper Palaeolithic peoples gradually adapting to the environmental changes. Some aspects of the two tool kits may corroborate this, particularly the bone and antler industry which is common to both (Morrison 1980). There is however a sharp differentiation of site location, the Upper Palaeolithic being virtually restricted to caves, while most Mesolithic sites are open, but this may merely represent an adaptation to the environmental changes (Binford 1968). Also, some examples of the continued use of rock shelters are available. Sheldon in the southern Pennines has an assemblage with both late Upper Palaeolithic and Early Mesolithic tool types (Radley 1968). Wetton Mill rock shelter in Staffordshire has no industry earlier than the Mesolithic, but this appears to be associated with a relict late glacial fauna of arctic fox, arctic hare and wolverine which exist alongside red deer and wild pig (Kelly 1976). On the whole, however, the Early Mesolithic industries are strongly associated with the initial invasion of woodland and its concomitant fauna across Britain (Jacobi 1978a).

Early Mesolithic sites are found in both upland and lowland Britain. The tool kits are relatively simple and similar in both locations. They are characterised by large and fairly crude microliths, principally obliquely-blunted points, with large triangles and long trapezes. These were probably used as projectile armatures (Barnes 1982), and give the generic name 'Broad Blade' (coined by Francis Buckley in 1924) to the industry. Endscrapers and burins are associated with these microliths. Bone and antler tools are relatively common, including spears which are barbed on one side only. It is this aspect of the technology which gives these Early Mesolithic sites their other name of 'Maglemosian' (Mellars 1974). Only four sites, all in northern Britain, have a slightly different industry (Star Carr, Flixton 1, Pointed Stone and Warcock Hill South) based on different types of flint (Mellars 1974, Jacobi 1978a, Switsur and Jacobi 1975). The explanation for this variant is unknown: neither a function of seasonality, as these sites are both highland and lowland, nor of chronology as they are contemporary with the more usual Early Mesolithic assemblages. The only other variation is the abundance of tranchet axes, which are commonly found in Southern England. It was thought that they were entirely absent from northern Britain (eg Clark 1932), but a small quantity have now been found in the Pennines (Radley 1964, Manby 1964, Davies 1964). Star Carr (Clark 1954) and Misterton Carr (Buckland and Dolby 1973) in the lowland to the east of the Pennines have produced a relatively large number of these axes.

The end of the Early Mesolithic occurred between 6800-6500 bc (Switsur and Jacobi 1975), with the appearance of Narrow Blade or 'geometric' microliths. The earliest dated Late Mesolithic site is Filpoke Beacon, near Hartlepool, at 6810 ± 140 bc, corresponding with pollen zone VIa (Switsur and Jacobi 1979), but the majority of the dates are after 6600 bc. In southern Britain, the Narrow Blade microliths were added to the Broad Blade repertoire, whereas in northern Britain, they replaced the Broad Blade microliths entirely. Axes disappeared from northern Britain and the Horsham points characteristic of south-east England also disappeared (Mellars 1974). These later Narrow Blade microliths appear to be projectile barbs which were hafted onto wooden shafts. Thus, they may have replaced the bone and antler spears of the Early Mesolithic (Mellars 1976a). They may also have been used to form a number of tools associated with gathering rather than hunting, such as sickles (Clarke 1976). The major advantage of using component parts rather than a single piece of raw material is the ability to easily curate tools. The evidence for increasing scarcity of good flint supplies through the Mesolithic (Pitts and Jacobi 1979) underlines the need for curation. Narrow Blade microliths were much smaller and more finely-worked than the Broad Blade forms, and occur in many forms, although three shapes predominate. Most common are the scalene triangles, found throughout Britain in both upland and lowland. They predominate throughout the Late Mesolithic. Secondly, some tool kits are dominated by rod-shaped microliths. These are found at upland sites only and may be, on the scanty evidence we have, a late development in the fourth millenium bc (Switsur and Jacobi 1975). Domination of assemblages by the third type of microlith, the rhomboid, is rare and represented only at three (undated) sites in the Pennines (Barnes 1982).

Apart from Scottish sites which may extend well into the third millennium bc (Mellars 1974), the end of the Mesolithic has been dated to between 3500-3000 bc. Dunford Bridge site B in the southern Pennines has a date of 3430 ± 80 bc (Radley <u>et al</u> 1974), and Wawcott in Berkshire at 3310 ± 130 bc (Froom 1974). Dating the different periods of the Mesolithic is still uncertain (Bonsall 1975) and it is possible that the sequence through the period may be extended in the future. Certainly, Radley (1969) thought that the Mesolithic industries overlapped with those of the Neolithic, thus lasting until 3000 bc at least, in both the Pennines and the North York Moors.

Generally, the Late Mesolithic is represented by many more sites than the earlier period. This is partly an artefact of its longer time span (three millennia compared to the two of the Early Mesolithic), but possibly also reflects a greater population density (Morrison 1980), perhaps as a result of increasing sea levels. Although at the Early/Late Mesolithic transition Britain was still attached to the Continent (Godwin <u>et al</u> 1958), the North Sea Plain was under water which greatly restricted the area available for exploitation (Jacobi 1973).

The Mesolithic in the Soyland Moor Area

The millstone grit area of the central Pennines has produced prolific Mesolithic flint sites since fieldwalkers first began to search in the 1870's (Hallam 1960). Soyland Moor is on the northern edge of the most concentrated area of sites (see Figure 1), but is

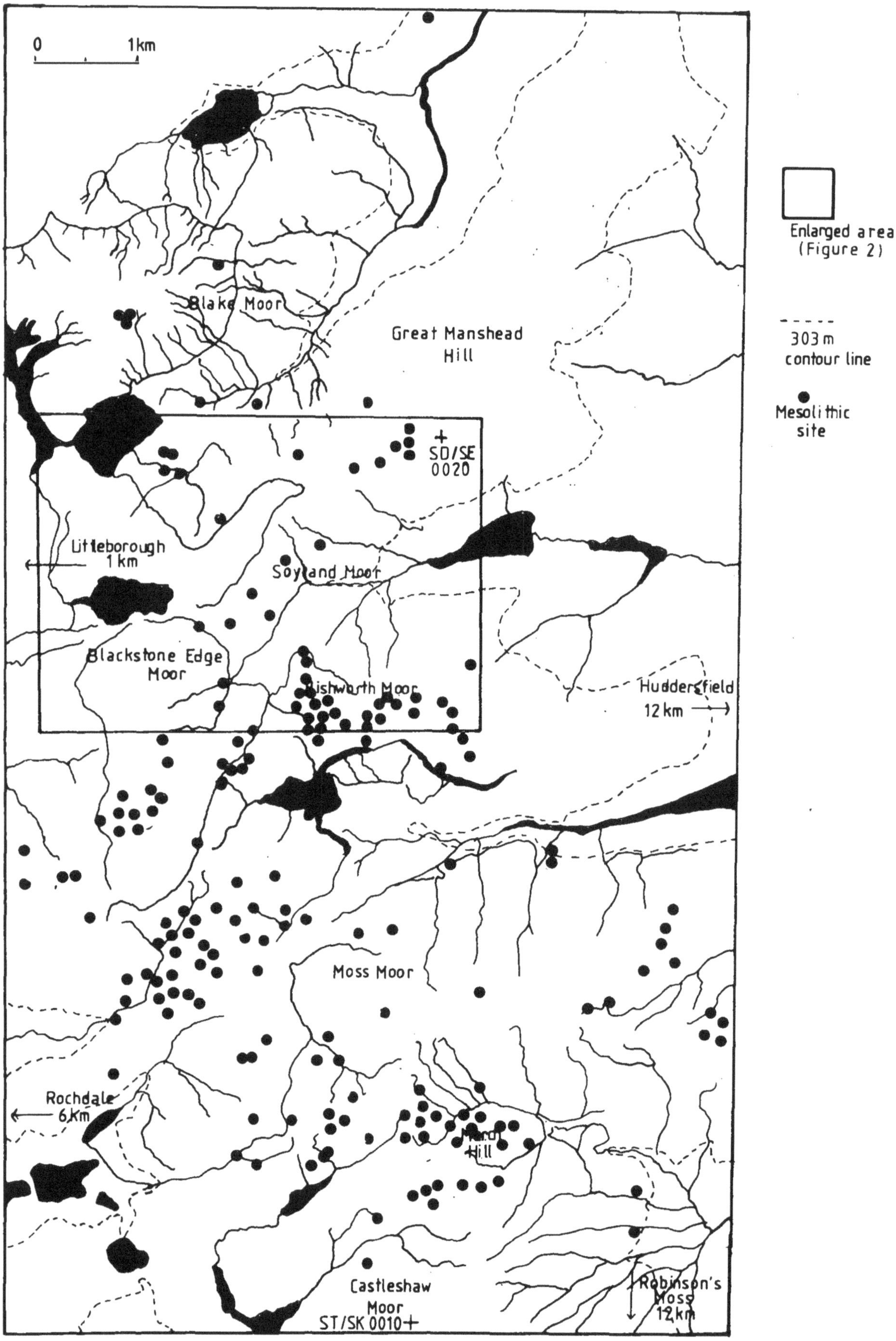

Figure 1 Location of major Mesolithic sites in the central Pennines Data from Wymer 1977.

surrounded by a great number of sites (see Figure 2). One of the major workers in the area was Francis Buckley, who located over 250 find spots in the Marsden area, bounded by Soyland Moor to the north, and Longdendale in the south, and excavated over twenty sites in the 1920's. His work provided the core of the Mesolithic evidence for the area. It was Buckley who first noticed the typological differences within the assemblages and named them Broad Blade and Narrow Blade. The lack of stratigraphy meant that the chronological sequence was indeterminable, and Buckley himself thought that the Broad Blade industry was later. Clark (1932) reversed this on typological grounds, a sequence which was corroborated by radiocarbon dating (Switsur and Jacobi 1975). Unfortunately Buckley was not as addicted to publication as he was to flint collecting. Petch (1924) summarised his findings and Buckley's original notebooks are available for inspection in the Tolson Memorial Museum in Huddersfield, but archaeologists are still trying to reconstruct many of his excavations. This task is not made any easier by Buckley's habit of donating flints to a large number of museums around the country. These somewhat haphazard methods, typical of his time, have unfortunately been perpetuated in this area, and even recent and well-excavated sites rarely see the light of publication, while flint assemblages are 'lost' in museums or remain in private collections. A good example of this is Ickornshaw Moor, an important site to the north-west of Soyland Moor. This site was well-excavated in the early 1960's and provided with radiocarbon dates and an associated pollen diagram but only the date has been published (Godwin and Willis 1964). Further south, the uplands above Sheffield have fared much better, and much of our present day understanding of the typology, Mesolithic dwellings and economy come from studies in this area (eg Radley et al 1974, Radley and Mellars 1964).

The large number of Mesolithic sites found in the central Pennines compared with other areas may well be due in part to the number of local field workers, and, as all the Pennine Mesolithic sites are found at the interface between mineral soil and peat, to the extent of peat erosion in the area. However, the concentration of sites is in part a reality (Manby 1964). The limestone areas to the north and south, while not devoid of sites, were apparently sparsely occupied, and millstone grit areas such as Kinder and Bleaklow show a similar sparsity of find spots (Radley 1964). The reasons for this concentration in the narrowest part of the Pennine chain are not understood. The central Pennines also constitutes the type area for the upland Mesolithic. Early Mesolithic sites are not numerous, but are scattered across the area. Most early assemblages are dominated by obliquely blunted points, and grey/white opaque flint usually constitutes 80-98% of the raw material. The earliest dated site is Lominot III, 8 km south of Soyland, at 7615 ± 350 bc (Switsur and Jacobi 1975). Warcock Hill South, 2 km to the south of Lominot, is the one exception, with an assemblage similar to Star Carr and Flixton Site 1, and its industry is composed mainly of translucent flint (Radley and Mellars 1964, 1978a). Its date of 7260 ± 340 bc (Switsur and Jacobi 1975) suggests it was at least broadly contemporary with the other Broad Blade sites (although the large standard deviation leaves room for doubt).

The earliest date for a central Pennine Narrow Blade assemblage is 6660 ± 110 bc for Warcock Hill site III (Switsur and Jacobi 1975).

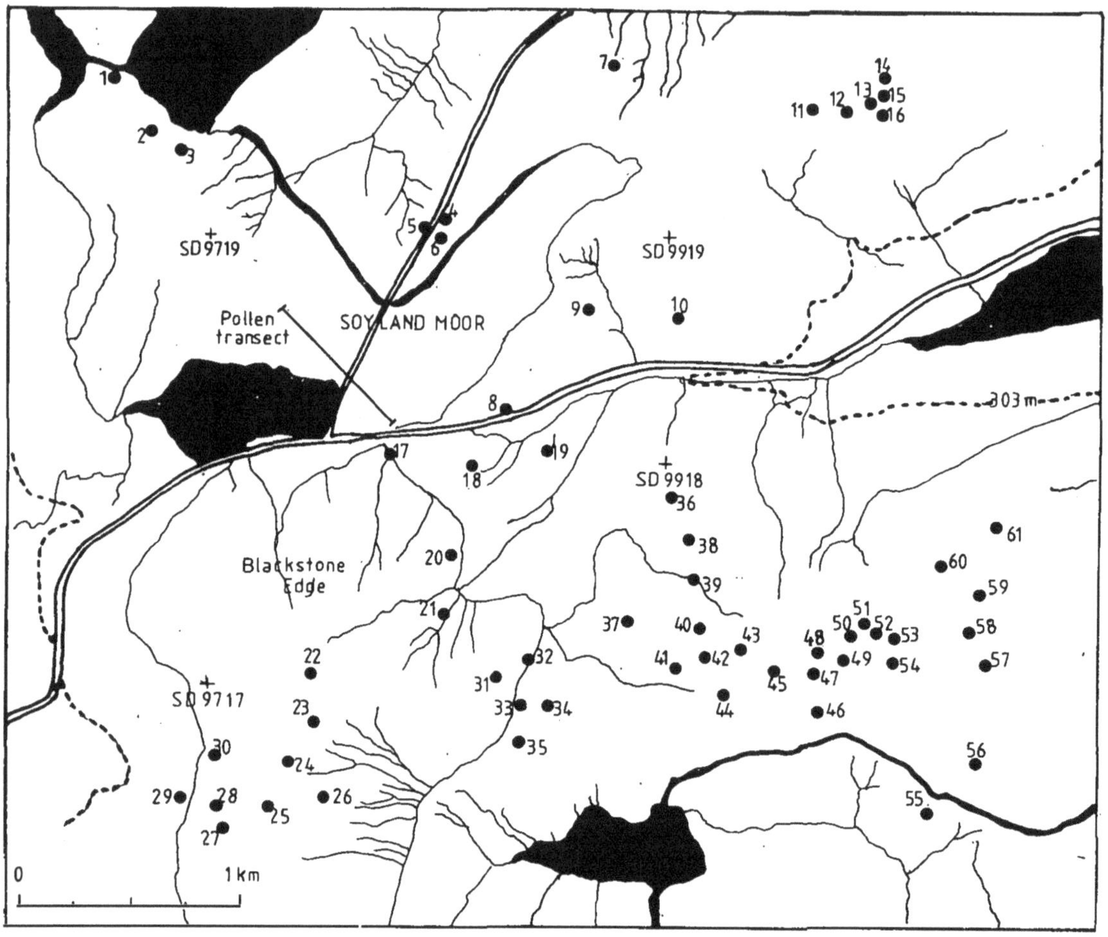

Figure 2 Location of Mesolithic find spots and sites in the vicinity of Soil and Moor

Note: Lack of published data prohibits the differentiation of Early and Late Mesolithic sites.

Unfortunately, there are large standard deviations for all the radiocarbon dates for the central Pennine sites, and the dates serve little purpose other than to underline the chronological sequence of Broad Blade to Narrow Blade. These later assemblages are usually dominated by the use of various types of translucent flint instead of the grey/white opaque material of the earlier period, and an increasing amount of chert occurs (Barnes 1982). Sites with assemblages dominated by all three major microlith forms are found in the central Pennines. Most assemblages are dominated by scalene triangles, including most of the sites around Soyland and Rishworth Moor (Petch 1924, Barnes 1982). Rod microliths dominate at Rocher Moss site 2, which is dated to 3880 ± 110 bc (Switsur and Jacobi 1975) and may therefore represent a late development. White Hill North and Red Ratcher are dominated by the rare rhomboid microlith. This only occurs elsewhere on Beeley Moor, Derbyshire (Barnes 1982) and, as yet, none are dated. The latest date we have for the Narrow Blade sites in the area is 3660 ± 120 bc at Lominot site IV, which is dominated by scalene triangles (Switsur and Jacobi 1975). Further dating may well extend the length of this period, which appears at some sites to overlap with the Neolithic (Barnes 1982).

Figure 2 Key to Mesolithic find spots and sites in the vicinity of Soyland Moor

Data from Wymer (1977)

1. White Holme Clough
2. Farther Hill
3. Farther Hill
4. Soyland Moor
5. Wicken Hill
6. Woolpack Hill, Soyland
7. Manshead Hill
8. Black Castle Hill
9. Knave Holes Clough
10. Black Castle Hill, Soyland Moor
11. Collin Hill
12. Manshead Hill
13. Rishworth, Man's Road
14. Top of Man's Head
15. Manshead End
16. Great Manshead Hill
17. Slate Pit Hill Site 2
18. Black Castle Clough
19. Waystone Edge
20. Dick Slack 2
21. Dick Slack
22. Letter Stones
23. Blackstone Edge
24. Blackstone Edge south-east
25. Robin Hood's Bay
26. Windy Hill
27. Blackstone Edge, Littleborough side
28. Blackstone Edge
29. Blackstone Edge (2 sites)
30. Blackstone Edge
31. Flint Hill (4 sites)
32. Rishworth Drain
33. Culvert 20, near Rishworth resevoir
34. Rishworth (2 sites; Drain 1, Drain 2)
35. Dick Slack 3
36. Warm Withens
37. Near Flint Hill
38. Flint Hill
39. Warm Withens Chert site
40. 2 sites: Catstones first peak, Dick Slack 1
41. Flint Hill
42. 2 sites: near Catstones main site and Flint Hill
43. 2 sites: Cat Hill and Catstones
44. 2 sites: Flint Hill and Rishworth Moor
45. White Hill
46. Catstones, above Green Withens resevoir
47. 2 sites: Green Withens and Flint Hill
48. Catstones site 5
49. Catstones Hill
50. Catstones Hill
51. West of Catstones Hill
52. Catstones
53. White Isles
54. Catstones site 4
55. Castle Dean, Rishworth
56. Joiner Stones
57. 2 sites: Dog Hill and Castle Dean
58. Dog Hill
59. Dog Hill
60. Dog Hill
61. Lench Hill, near disused quarry

The location of all the Mesolithic sites in the area is remarkably uniform. Firstly, all sites are found at the interface between the mineral soil and the peat (Mellars 1974). The majority of the sites occur between 350-500 m (Jacobi 1973). This is not a factor of peat erosion, which extends down to 300 m and is most severe above 500 m (Jacobi et al 1976). The areas of sandstone that are interspersed with the millstone grit were strongly favoured and hollows were avoided, suggesting that well-drained areas were a priority when choosing a site (Petch 1924). Most sites have a southerly aspect and are located near the tops of ridges and hills, giving good views over the surrounding plateau and nearby valleys (Watson 1952), but allowing some protection from exposure. Two types of sites are found. The majority are small (10-15 sq m) discrete units of flints (Petch 1924, Mellars 1976a), but some sites consist of a large area of scattered flints with a number of more concentrated patches within it (Jacobi 1978a). In the Soyland area, Blackstone Edge is a good example of the latter type, while the Rishworth sites to the south, and those to the north of Soyland, are discrete units (Petch 1924, Buckley notebooks).

The explanation of the use of these sites is based upon the concept of seasonal movements of the Mesolithic groups, to the uplands in summer and the lowlands in winter (Morrison 1980, Mellars 1976a). According to this model, winter would be spent away from the high ground to avoid the harsher climate and concomitant shorter growing season, whereas summer would be spent in areas with lighter tree cover, as this would allow the greater production of edible understory herbs and hazel nuts to be exploited (dense forest in the lowlands would suppress understorey growth). These factors would also apply to the herbivores who would also migrate to higher ground to minimise insect irritation (Jacobi 1978a). The movements of man may well be closely associated with that of the herbivores (Mellars 1976a), thus giving greater weight to the annual round hypothesis. In the Early Mesolithic at least, the major winter settlement of these Pennine groups appears to have been the Lincolnshire Edge which lies parallel to the central Pennines (Jacobi 1978a). The sites there are larger, typical of winter settlements (Mellars 1976a), and the ubiquitous grey/white opaque flint used in the Early Mesolithic in the uplands also originates from this area (Jacobi 1978a). The upland sites in the Early Mesolithic may represent base camps which are used throughout the summer season, as they are relatively large and contained balanced assemblages of hunting and processing equipment (Mellars 1976a). In the Late Mesolithic, most sites appear to be more ephemeral: they are smaller in area, with a lesser quantity and range of tools. These tools are mainly microliths, possibly indicating that hunting was the major activity with little domestic or maintenance activity (Mellars 1976a).

The simple model of uplands in summer and lowlands in winter is complicated by a number of factors. The existence of 'intermediate' sites in both the Early and Late Mesolithic, at c 150 m OD on the flanks of the Pennines has to be explained. While these sites are not found on the central Pennine flanks themselves, there is enough evidence from other areas to suggest that this may be merely a factor of sampling. To the south, the sites of Deepcar (Radley and Mellars 1964) and Unstone (Courtney 1978) are in this intermediate position. Upleatham, on the flanks of the Cleveland Hills at 165 m OD (Spratt et *al* 1976) is similar. All these sites are large, and Deepcar produced evidence of a well constructed shelter. These sites have been interpreted as base camps from which hunting parties left to exploit the high summer ranges. Another complicating factor is the second Late Mesolithic type of site found on the high Pennines, such as those found at Blackstone Edge when small discrete sites are found within a larger utilised area. It may be that the number of this type of site has been underestimated (Jacobi 1978a). Whatever the interpretation, there is enough evidence to suggest that small and simple 'overnight' camps are not the only type of summer site to be found, in either the Early or Late Mesolithic.

While the data suggest seasonal movement, this may not have occurred each year. In mild winters, herbivores may well have remained in the summer ranges (Mellars 1976b, Jacobi 1978a), and man would perforce have to remain also, as meat would constitute a large part of the winter diet (Mellars 1976a). The agglomerations of upland sites could then represent winter base camps, as both archaeological and ethnographic data point to larger settlements in winter (Mellars 1976a).

Direct evidence for the economy of the upland Mesolithic groups is virtually lacking as organic material is not preserved in the acid soils and peats (apart from carbonised hazel nuts). Extrapolating from lowland sites, it has been suggested that red deer constituted the major meat resource throughout the Mesolithic (Mellars 1975), and that a wide range of vegetable food was gathered. Hazel nuts have been regarded as the staple vegetable resource due to the occurrence of carbonised hazel nut shells in a number of Mesolithic hearths (Mellars 1976a), and the prevalence of hazel in postglacial pollen diagrams from the uplands (eg Smith 1970). This evidence can occasionally be taken too far: Jacobi (1978b) estimates the Mesolithic population based on the food potential of the hazel harvest, as calculated by hazel pollen values, which is a tenuous and not very profitable exercise. The view of Mesolithic man as a simple hunter-gatherer (eg Iversen 1949) is however slowly losing ground, not least because ethnographical studies show that there is no such thing as random hunter-gathering (eg Lee and DeVore 1968). Exactly what economic strategies Mesolithic man was pursuing in the uplands will remain speculative, due to the dearth of direct evidence. The burning hypothesis put forward by Simmons (1969) and Mellars (1976b) is the first step in evolving a more sophisticated set of models of the economic behaviour of mobile groups in the Mesolithic period. The evidence from the Soyland Moor pollen diagrams is used to test these models, and to study the interaction between the economy and the society, and the role these economic strategies played in the change to the Neolithic.

CHAPTER 3

METHODS OF STUDY

Sampling Strategy

A series of four cores was taken across Soyland Moor from the extreme southern edge of the mire to the deepest area of peat. The reasons for taking four cores were:

1. To provide a full local history of the mire showing rates of peat growth and spread.

2. To provide data on the direction and geographical extent of the vegetation.

3. Evidence for clearance is often highly localised, and may be seen in one area of the mire and not another.

4. Small, statistically insignificant changes in pollen profiles could be verified as representing 'real' changes, or disproved, by comparison with the other diagrams.

Figure 3 represents the section through the deepest peat deposits, showing the topography of the original shallow basin. Figure 4 shows the position of the section and the subsequent positions of the four cores taken approximately 100 m from each other. Site A was taken as near to the southern edge of the mire as possible. The position of the A58 in between Blackstone Edge and Soyland Moor precluded the sampling of the extreme edge of the mire, and trial cores were taken at Site A to ensure that deposits were not disturbed by the road construction. The four series of cores were obtained with a Russian borer. The 50 cm cores were taken alternately from two close points, avoiding disturbance by the cutting point of the borer below the core chamber. At site D, sampling was difficult due to areas of very wet peat at differing levels below the surface which would not form cores. Two series of cores were taken 2 m apart, avoiding these very wet areas. The two diagrams subsequently constructed were married at 596 cm below the surface. Below the peat was a weathered sandy substrata with little or no organic content, which was sampled to provide information of the immediate pre-peat vegetation. Sampling here was undertaken with the screw of a Hiller borer, the only implement which would penetrate the material. Possible problems of contamination are discussed later.

The peat cores were placed in plastic guttering and separately wrapped in polythene on site. These were stored at 5° C until they were prepared for counting. The pre-peat material was carefully surface-cleaned and small amounts were placed in labelled tubes on site, as the material would not travel intact. Just prior to preparation, the peat cores were cut into centimetre slices. While it is preferable to surface-clean a core and remove only the peat to be prepared (Moore and Webb 1978), the close sampling interval employed made this technique impossible. The peat for preparation was taken from the centre of each slice. The remaining peat was stored in separate sealed and labelled polythene bags at 5° C.

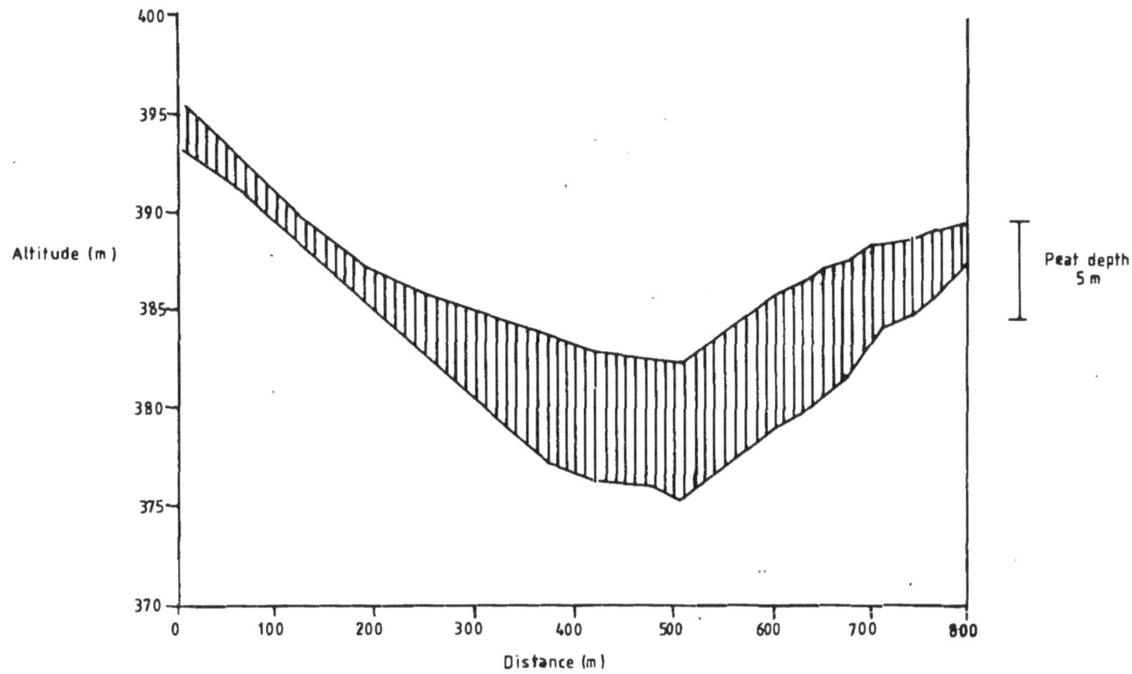

Figure 3 Profile of the shallow valley at Soyland Moor, indicating the basal topography and depth of peat cover

Note: The profile is exaggerated by the differing scales of the two axes.
Peat deposit indicated by shading.

Pollen Analysis

The preparation of the peat samples for counting followed the standard procedure of potassium hydroxide digestion followed by acetolysis (eg Moore and Webb 1978). A safranin stain and glycerol jelly mounting medium were used, and three slides were made up from each sample. Appendix I gives full details of the procedure.

The microscope used for all the pollen counting was the Wild M20, fitted with phase contrast condenser and objectives, as recommended by Moore and Webb (1978). The majority of the counting was done under x 400 magnification, with x 1000 oil immersion used where necessary for identification.

As this research was only dealing with the earlier part of each site sequence (up to the Elm Decline), a close sampling interval was possible. Counting was carried out at every centimetre, to give a close look at the changes in the vegetation. Only in the upper part of SMD was counting at a greater interval (5 cm). The faster rate of peat growth at SMD meant that this sampling interval was reasonably similar to SMC, and time also precluded closer sampling. The strategy employed for counting pollen grains was aimed at producing a representative and reproducible series of pollen counts. Non-random pollen distribution on the slide was counteracted by traversing both the edges and the central part of the slides (Brookes and Thomas 1967). At least six traverses of each slide were made. To counteract the possibility of inadequate mixing of the glycerol jelly with the pollen, all three slides from each sample were examined. Faegri

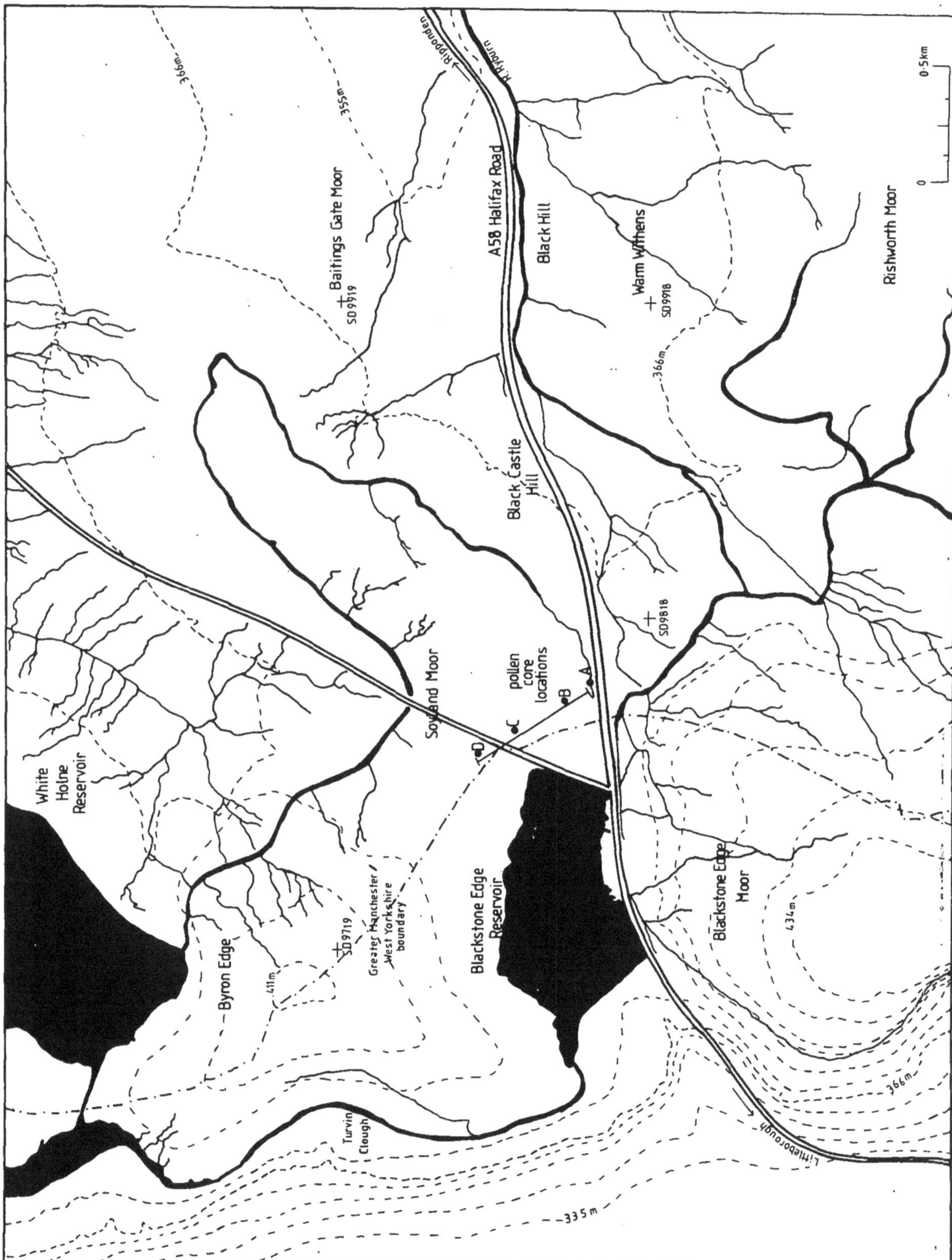

Figure 4 Location of the pollen sites and the line of the section through the mire

(1951) points to a problem of over-representation of large grains on the edges of the field of vision. Thus, only grains represented by 50% or more on the edge of the field of vision were counted. The problem of crumpled, broken and degraded grains in the samples was to some extent insuperable. Some grains are readily identifiable even if very degraded or represented only by tiny fragments (eg *Ulmus*). Also, some grains degrade more rapidly than other (eg *Populus*) while others fall apart easily (eg *Calluna* tetrads). Unidentifiable remains were not included in the count, as it was not always possible to identify them as pollen. Over-representation due to counting of identifiable fragments was corrected by counting on a minimum numbers basis. Thus, for example, for one grain of *Pinus* to be included in the count, two bladders and a central body had to be present, and fragments of *Pinus* were collated in this way. 50% of a grain, or a component part, had to be present before being allowed to appear in the count. These rules applied to all except rare herb pollen types, where any fragment, if it could be positively identified, was counted.

When analysing the pollen count, the frequency of each pollen type was constructed as a percentage of a fixed sum. While absolute diagrams may provide the ideal way of looking at pollen data, the extra expense and time involved (Birks and Birks 1980) precluded their use here. To construct a percentage diagram it is necessary to decide, firstly, what pollen types are to be included in the sum, and, secondly, how many grains constitute that sum at each level. In early pollen work, where the major aim was the history of woodland development, this sum included tree pollen only (Wright and Patten 1963). This study, which was interested in the extent of the woodland, the mire vegetation and the activities of man, required a sum that was representative of the entire vegetation (Faegri 1956). Therefore, the sum included all pollen types. It was also felt that a total pollen sum avoided any assumptions about the status of the vegetation types (Wright and Patten 1963), such as those made by Janssen (1959), when he excluded 'local' pollen. *Sphagnum* spores alone were excluded from the sum, as they are produced in an erratic fashion, occasionally swamping the pollen by a factor of 100 or more, and do not accurately reflect the extent of *Sphagnum* in the landscape. The major drawback of the use of a total pollen sum was the difficulty in comparing the Soyland diagrams with other published diagrams, which still tend to use sums based on tree pollen only. Summary diagrams of the relative composition of trees, shrubs, dwarf shrubs and herbs were constructed partly to enable gross but direct comparison with other sites. In the summary diagrams, *Corylus* is classed as a shrub, in accordance with tradition. This facilitated comparison with other sites, although the distinction between trees and shrubs is mostly academic in the highlands. The use of a total pollen sum in the main diagrams precluded the problem of classifying *Corylus* one way or another.

The choice of the number of pollen grains which constituted the sum was a compromise between the need for as large a sum as possible and the practical feasibility of counting in the time available. As small-scale vegetation changes, especially clearances, were high on the list of priorities, a high count was necessary to smooth out statistical variations (Moore and Webb 1978). Herb pollen is important for indications of culture, but their low pollen productivity also necessitates a high count (Vuorela 1977). A further consideration was the comparison of contemporary levels in all four diagrams, which would rely on the fluctuations to be a result of changes in the vegetation, rather than vagaries in the pollen count itself. The larger the sum, the more confidence could be placed in the results. Birks and Birks (1980) point out that a sum of 300-500 grains is statistically significant, that is, replicable, while Moore and Webb (1978) state that a sum of 1000 is optimal. At Soyland, the actual number of grains counted per level was 800 (plus any *Sphagnum*) spores). There were two circumstances under which this number was changed. In some levels, one type would predominate to such an extent that other pollen types were negligible. Instead of excluding the dominant type from the sum, the count was raised (up to 1,400 grains maximum). Secondly, pollen was occasionally very sparse, and a count of 800 was not possible from the three slides prepared. In these cases, all the grains were counted and this figure used as the pollen sum. The minimum count was 500 grains, below which further samples were prepared and counted.

Pollen keys were the preliminary tool used for identification of the pollen counted, in particular that of Moore and Webb (1978). Verification was supplied by the use of type slides. Where further verification was needed, or type slides were not available, identification was made by Dr P D Moore of King's College, University of London. Appendix II lists all the pollen types found at Soyland Moor, and which species are included in each. Appendix III lists the specific characteristics of the more unusual pollen types encountered.

The Diagram and its Interpretation

All four diagrams, which are sited at the very end of the book for easy reference, were presented in a similar format, based on histograms for the percentages of each species through time as represented by the stratigraphy. This follows standard procedure (eg Faegri and Iversen 1975). The data are presented in a raw form, with no exaggerated profiles and no correction factors for differential pollen production.

The division of the diagrams into zones was based on gross changes in the AP and *Corylus* values, particularly at the rational limit or decline of a pollen type. The 'rational limit' is the term used by Smith and Pilcher (1973) to denote the point at which a pollen profile registers the beginning of a rise to sustained high values, usually above 5% TP. (The term 'empirical limit' is also used when discussing the pollen data and refers to the point when a type is first consistently present.) Zonation was carried out on each diagram separately, with zone numbers starting at the base of each diagram. Thus there is no inherent synchroneity implied by the zoning across the four diagrams, although this was often the case.

The use of local pollen zones, rather than the standard zonations of Godwin or Blytt and Sernander (Godwin 1975a), reflects the differences within the vegetation sequence at Soyland to that of other areas. Table 1 shows the broad correlation of the two standard zonation schemes with the local Soyland Moor scheme.

TABLE 1: Postglacial Zonation Correlations

Godwin zonation with generalised dates (Godwin 1975a)	Blytt-Sernander Sequence (Godwin 1975a)	Equivalent SMC Zones with dates
Zone IV 8300-7600 bc	Pre-boreal	(SMD 1, SMC 1)
Zone V 7600-6200 bc	Boreal	SMC 2 6700-6160 bc
Zone VI 6200-5500 bc	Boreal	SMC 3 6160-5690 bc
		SMC 4 5690-4160 bc
Zone VIIa 5500-3000 bc	Atlantic	SMC 5 4160-2615 bc
VIIa/VIIb boundary (Elm Decline) 3000 bc	Atlantic/Sub-boreal transition	SMC 5/6 2615 bc

The first problem confronting the palynologist when interpreting the four Soyland pollen diagrams is the use of percentage format, which can produce fluctuations which bear no relationship to real changes in the pollen input (Moore and Webb 1978). The problem is to decide which feature is changing in reality, and which feature changes merely as a result of changes in percentages. This is not, however, a major problem, as comparisons of absolute and percentage diagrams show similar patterns (Birks and Birks 1980). It is the large pollen producers only which are affected, and a real increase or decrease in one pollen type tends to affect the rest of the pollen spectrum, rather than just one other type. In the Soyland diagrams, the use of a TP sum minimises the problems by spreading the percentage fluctuations over the whole range of pollen types, rather than just the trees.

There are three major tools used in the interpretation of pollen diagrams. The ecological associations in modern habitats are a major guide to the understanding of the past environment. This can be misleading as plant associations are not necessarily stable (Watts 1973). Even today, their variations are wide, and in the past the differing competition conditions could produce very different associations. The appearance of indicator species has also been used for such parameters as temperature and soil fertility. This assumes that their ecological responses are the same as today's, and also that we have adequate autecological knowledge of the species (Moore 1980). The problem of differing ecotypes within the same species is particularly problematic. The identification of pollen to species level is necessary for their use as indicators, and in many cases this is not possible. However these approaches can be used with caution, ie, as long as groups of interacting taxa are used, and any contradictory evidence adequately explained (Birks and Birks 1980).

Modern pollen production studies are now providing us with some direct knowledge to translate pollen values into the position of the species in the landscape itself. Although problems arise and many qualifications have to be made, it is proving to be the soundest approach to interpretation (Birks and Birks 1980). Pollen production is different for each species, but the production figures for each species can themselves change for a variety of reasons. Many factors affect the day-to-day production, particularly the microclimate and the inter-relationships with the surrounding vegetation (Tinsley 1972). However, such fluctuations tend to be smoothed out over time (Mandrioli et al 1980). Long term climatic changes (Birks and Birks 1980) and altitudinal and latitudinal differences (Webb et al 1978) have more important effects on pollen production, some of which are difficult to calculate.

Initially, much of the pollen production research was done within forests (eg Andersen 1967), which was of little direct use in interpreting data from mires and lakes. Today, studies have been made in many differing environments, including fen carr (Caseldine and Gordon 1978), lakes (Prentice 1978), mires (Beckett 1979) and across open woodland/heathland transitions (Tinsley and Smith 1974). These provide useful information which can be applied to different areas, given care. The dispersal rate of pollen also greatly affects representation in pollen diagrams. As this had a direct bearing on the spatial analysis of the Soyland Moor data, this is discussed in Chapter 5.

When interpreting an individual site, one of the major methods of evaluating the environmental data is comparison with other similar sites. Care has to be taken to avoid circular arguments, and local factors have to be taken into account. At Soyland, most of the sites used for comparative purposes are Pennine or other highland sites, often the North York Moors. Lowland sites were used when tracing altitudinal differences. There are many pollen studies in the southern and central Pennines, but unfortunately, most of these sites have late peat initiation, usually the end of Zone VI or during VIIa (Tallis 1964a). Thus, their use for comparative purposes was limited. A few Pennine sites have earlier peat accumulation, especially in the northern Pennines (eg Turner and Hodgson 1979). Nearer to the Soyland area is Fountain's Earth in Nidderdale (Tinsley 1975), where peat began to accumulate in early Zone VI. Tarn Moss at Malham (Pigott and Pigott 1963) covers the whole of the late glacial and postglacial, but its position on the limestone shows a different type of vegetation. The best site for direct comparison with Soyland was Robinson's Moss in Longdendale, 20 km to the south (Tallis unpublished). Robinson's Moss is similarly situated on the millstone grit, and has early peat accumulation.

Despite a number of seemingly insurmountable problems in assessing pollen data, replicable and ecologically sound results are usually produced, indicating, at the very least, that pollen representation is standard enough over time and space to allow interpretation to be made. To point out the problems is to pave the way to solutions, not to destroy the validity of the method.

Radiocarbon Dating

Some of the unusual features of the Soyland diagram needed to be absolutely dated before interpretation was possible. Dr V R Switsur of the Sub-Department of Quaternary Research at the University of Cambridge generously undertook to produce a series of dates from peat material from the original cores. Appendix V by Dr Switsur, gives a full technical discussion of the production of these eleven dates. Generally, the dates are based on a half-life of 5568, and are uncalibrated. All the dates referred to in this volume are in

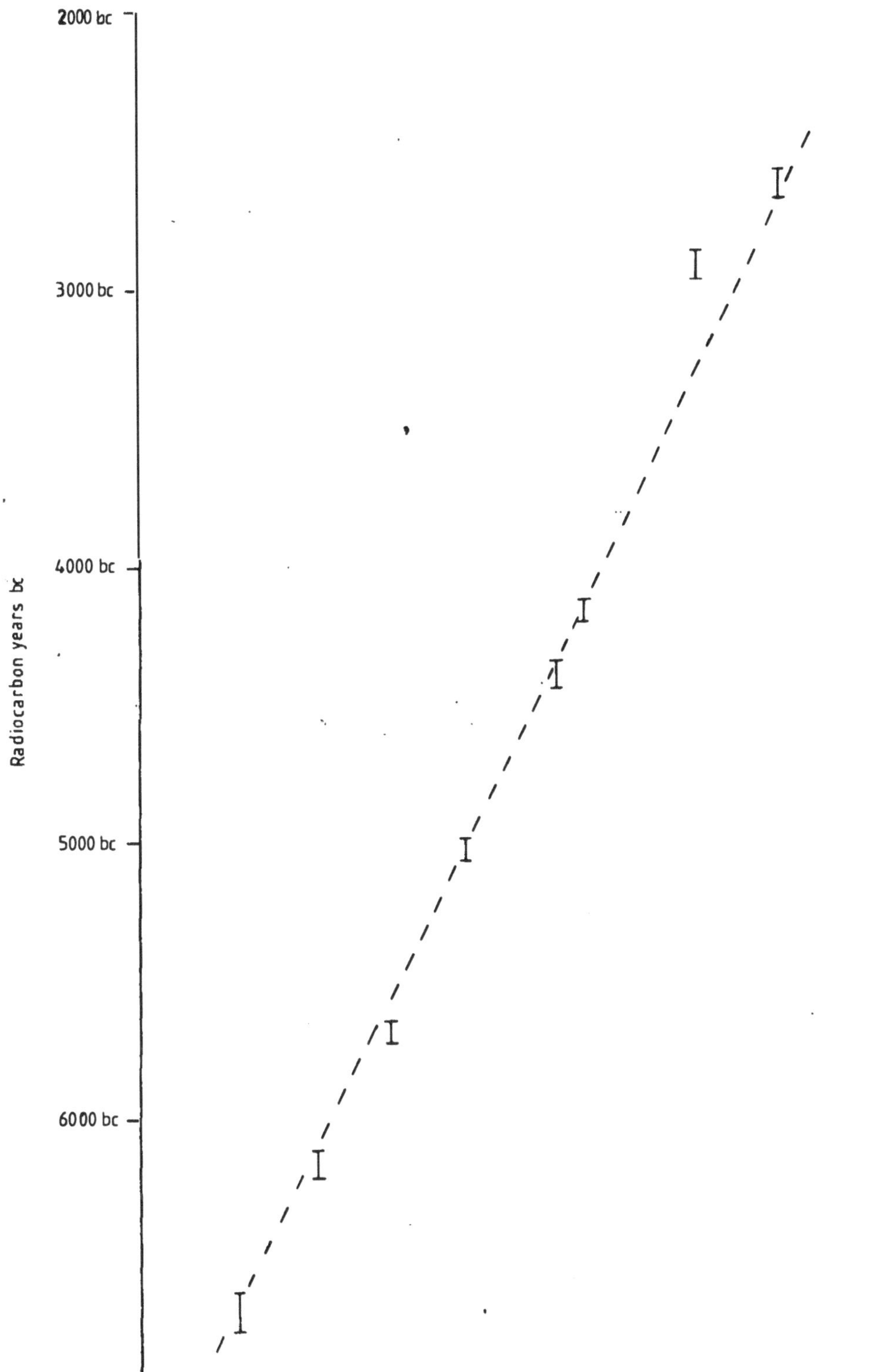

Figure 5 Soyland Moor peat growth rate as ascertained from the radiocarbon dates

radiocarbon years, and are thus designated by the small-case bc.

Soyland Moor Site C was provided with a series of eight dates, from the base of the profile to the Elm Decline. These dates are as follows:

SMC 470 cm : Q 2385 4565 ± 50 bp (2615 bc)

SMC 500 cm : Q 2386 4865 ± 50 bp (2915 bc)

SMC 541 cm : Q 2387 6110 ± 40 bp (4160 bc)

SMC 551 cm : Q 2388 6340 ± 50 bp (4390 bc)

SMC 583 cm : Q 2389 6975 ± 40 bp (5025 bc)

SMC 609 cm : Q 2390 7640 ± 40 bp (5690 bc)

SMC 636 cm : Q 2391 8110 ± 50 bp (6160 bc)

SMC 664 cm : Q 2392 8650 ± 75 bp (6700 bc)

This series of eight dates provides one of the best sequences of pre-Elm Decline dates that we have in Britain, and allows comparison with other well-dated sites, such as Red Moss in Lancashire (Hibbert, Switsur and West 1971). The series also dates a small clearance at SMC 500 exclusive to this site. Figure 5 plots these dates, and shows that peat accumulation rates were almost uniform up to the Elm Decline. This is useful, as it allows extrapolation between the dates, based on 1 cm representing 20 years.

Soyland Moor Site D was provided with three dates; the base of the profile, the Elm Decline, and a date for the clearance phase with cereal grains at SMD 500 cm. This latter date was necessary to understand the relationship between the two clearances at SMC 500 and SMD 500, and to date what appeared to be a very early appearance of cereals. In some ways, SMD should have been the site to provide the full sequence of dates. It appeared to be older than SMC on the pollen evidence, and the Elm Decline appeared at a much higher level in the peat stratigraphy, thus enabling greater resolution. However, the problems caused by the very wet deposits and the subsequent matching of two sets of cores to produce the SMD diagram made it inherently less reliable. The three dates obtained are:

SMD 365 cm : Q 2393 3400 ± 100 bp (1450 bc)

SMD 500 cm : Q 2394 5820 ± 95 bp (3870 bc)

SMD 664 cm : Q 2395 8190 ± 70 bp (6240 bc)

The significance and correlations of these dates, and any possible problems of acceptance, are discussed with the pollen data in later chapters.

CHAPTER 4

POLLEN DIAGRAM DESCRIPTION, INTERPRETATION AND CORRELATION

In this chapter, the pollen data from the four sites sampled at Soyland Moor are discussed. The pollen diagrams themselves are located for easy reference after the references and appendices. The emphasis of this chapter is directed towards reconstruction of vegetation changes from the early postglacial to the Elm Decline and how Soyland Moor compares with other British sites. Soyland Moor site C (SMC) is used as the basis for interpreting the vegetation changes. As discussed in the last chapter, Soyland Moor Site D (SMD) is less suitable as the foundation stone, while the two other sites cover only the very end of the period concerned. The next chapter will concern itself with the very local vegetation changes that can be seen when comparing the four diagrams in detail. Here, they are compared only to confirm the position of Soyland Moor within the British sequence. Similarly, the description and discussion of both the Elm Decline and the evidence for human disturbance of the vegetation have been left until later chapters, where detailed description is necessary to evaluate the evidence.

Thus, SMC is described and interpreted in detail, while SMD, SMA and SMB are discussed in relation to SMC only. The pre-peat material (Zone 1 in all the diagrams, despite their differing ages), however, is discussed separately in each case, due to possible problems of contamination. Due to the depth of peat and hardness of this sandy pre-peat matrix, sampling had to be undertaken with the basal screw of a Hiller borer. Care was taken to ensure as little contamination as possible, but an open sampler tool cannot guarantee pure material. Downwashing of pollen within the sandy, large-particled matrix may also result in mixing of the chronological sequence (Dimbleby 1961b). Thus the pre-peat material is less reliable than the peat itself as an indicator of past vegetation. The problems of possible contamination are discussed in each case, although only in SMB did it appear to be an overriding factor.

Both the botanical and common names of plant species are used in the description and interpretation of the pollen diagrams. For the archaeological reader, Appendix II gives a list of species occurring in the diagrams with both forms of their names. It also lists the species included within a particular pollen morphological 'type'.

Soyland Moor Site C

This is the third core in the series, taken on a straight line with cores A and B, and at a hundred metres from B (see Map 3). The peat at SMC measured 664 cm in depth, and touched on the deepest part of the basin. The sandy pre-peat matrix was sampled with the screw of the Hiller borer to a depth of 690 cm, and nine samples were taken. The peat stratigraphy showed a highly amorphous character at the base, and the components of the peat were unidentifiable. Above 600 cm, *Eriophorum vaginatum* dominated the peat components, and was identifiable from well-preserved leaf-bases. If *Sphagnum* was an important component, as is suggested by its spore profile, it was too degraded to be identifiable by macro-analysis.

SMC 1 690-668 cm Pre-Peat Material

This pre-peat profile can be divided tentatively into two sections, splitting at 675 cm below the surface. However, evidence of contamination made it appropriate to discuss this data as a whole and to give it less consideration and weight than the data from the peat above. This contamination is evidenced by the incidence of *Fagus*, which did not occur in Britain until the mid-postglacial, and the appearance of *Quercus* at the very base, before *Ulmus*, an inversion of the normal migration (Godwin 1975a). Later in SMC 1, *Quercus* disappears, not to reappear until SMC 2, after the *Ulmus* curve is established. A few grains of *Alnus* appear at the base of SMC 1. This may be evidence for contamination, but they could also be explained as long distance transport of pollen from early *Alnus* stands, which established themselves in suitable habits in Britain in early Zone IV, usually to disappear again (Scaife 1982).

Description

Betula dominates the spectrum at the beginning of SMC 1 with values of nearly 50% of the total pollen (TP). Its values decline steadily through this zone to 10% TP at the SMC 1/2 boundary. *Corylus* begins at under 5% and remains low until 674 cm when values increase enormously to over 50% TP, displacing *Betula*. Pine is present, showing an erratic curve, but always above 5% TP. *Salix* values start at 5% but decline to 2% by the end of the zone. *Calluna* values are low at 5% TP, but increase slightly through the zone. The other ericales dominate the dwarf shrubs, rising to nearly 30% TP by 677 cm, then dropping to 5% TP. *Empetrum* follows the same pattern, with lower values. Both Gramineae and the Cyperaceae have high values at the beginning (reaching 17% and 12% respectively), but decline, ending the zone at under 5% TP. There is a wide range of herbs, often with high values. *Filipendula* and Umbelliferae show continuous curves, and the Composite types, Cruciferae and *Urtica dioica* are also abundant. The Polypodiaceae begin the zone at 15%, but decline to 2% by the end of the zone. *Pteridium aquilinum* is present, but in low numbers.

Interpretation and Correlations

Despite the slight evidence for contamination, the data show the vegetation patterns of the very early postglacial (Godwin's late Zone IV and early Zone V, dating from 8000 bc). The base of the profile to 674 cm represents Zone IV and shows tree birch (see Appendix III) as the dominant species. The amount of trees these high percentages represent is difficult to judge. In an open situation, *Betula* pollen may be produced in high quantities and also may travel much further than in closed woodland (Smith 1972). With such large percentages however, it is possible that birch was growing in the immediate vicinity, if not on the actual SMC site itself, from which position it would also dominate the diagram. Pine occurs at over 5% TP, which suggests local presence within birch woodland (O'Sullivan 1973). The presence of oak and elm pollen may be contamination or long distance transport from lower ground to the south west and south east.

The low hazel values at the base of SMC 1 suggest that, if it is in the area at all, it is in very low numbers. Hazel is a high pollen producer (Mandrioli *et al* 1980), except in shaded situations, where its vigour and flowering can be severely curtailed (Dimbleby 1978, Deacon 1974). The low values here cannot be attributed to shade and reflect the true position of hazel in the landscape. These low values effectively place this part of SMC 1 in Zone IV, the initial period of vegetation response to the postglacial warming. Willow pollen values are high at the beginning of the zone, and its percentages should be doubled at least for them to reflect its status in the vegetation, as willow is an insect-pollinated, low pollen-producing genus (Godwin 1975a). In this area, willow may be part of a select late glacial community, where various willow species would grow as bushes on open heath and grassland, rather than confined to wetter habitats as in later periods (Godwin 1975a). Heathland species are abundant, comprising Ericales type (*Vaccinium* or *Erica*) with a little *Calluna*. Taken into consideration with the occurrence of *Ulex* type, and the still high values of grasses and sedges, it is clear that much of the ground was open at this time, and thus strengthens the hypothesis that birch was not a dominant feature of the landscape, but may have been growing on site.

The herb communities as shown by the pollen underline the continuation of many late glacial features in these uplands despite the increased warmth and the migration of birch and pine into the immediate vicinity. In the higher areas of the Pennines (over 545 m OD), this relict late glacial flora could survive as late as 6500 bc (Tallis 1964a). The herb species are more numerous and more varied than during later zones, and many of the typical late glacial communities may well be represented. Some pollen types suggest the occurrence of a tall herb community, with *Filipendula* achieving 2% TP, *Rumex* and *Oxyria* types abundant, including *Rumex* cf *crispus* and cf *Oxyria digyna*, both indicators of the late glacial, or alpine situations (see Appendix III for identification of the latter two species). Umbelliferae and *Urtica dioica* are also common. Single occurrences of *Scabiosa columbaria*, *Valeriana officinalis* and *Geum* show the variety of tall herbs in this period. The occurrence of *Lycopodium alpinum*, *Artemisia*, *Lotus*, *Rumex acetosa* and *Empetrum* may indicate continuing late snow beds (Godwin 1975a). Cruciferae, Chenopodiaceae, *Melampyrum*, *Galium* and a large range of Compositae types are all indicative of disturbed ground. Godwin (1960) suggests that the presence of these herbs in the early postglacial point to continuing frost disturbance of soils. Other herbs merely indicate the existence of tracts of open ground, for example, *Ranunculus acris* type. The only woodland indicator present is a single occurrence of *Anemone*. The Polypodiaceae do, of course, include ferns which grow in woodland, but the unusually high percentages again suggest a throwback to the late glacial; for example, Blelham Bog in the Lake District (Pennington and Bonny 1970), records high Polypodiaceae only in Zone III of the late glacial. The large percentages of *Sphagnum* suggest considerable areas of waterlogged soils. Peat had not yet begun to form on this site, but conditions suitable for its initiation would appear to have been present.

The landscape at this period shows a marked similarity to that of Zone III of the late glacial, only the presence of birch and pine pointing to the later date of between 8350 and 7600 bc. The time lag may be explained simply in terms of altitude, where wooded areas would have been less extensive and more open. Thus conditions would have been favourable for heathland, tall herb and grass communities for a longer period. Connolly (1959) suggests that this extension may also be accounted for by low grazing pressure, allowing the typical late glacial communities to remain until deciduous forest shaded them out. Through Britain as a whole, the vegetation patterns show a high degree of variability from open conditions through to closed birch, or birch and pine forest. This can be seen as differing local responses to a common climatic amelioration, responses which are not entirely dependent on the migration rates of birch and pine. However, some of the variability is illusory, depending on how the tree percentages are interpreted: Thus, the 50% arboreal pollen (AP) at Soyland Moor should suggest a complete, but not closed, woodland cover (Tinsley and Smith 1974, Litchi-Federovich and Ritchie 1965). However, as previously discussed, the absence of any other major pollen producer is magnifying the position of birch in the landscape, but unfortunately, the factor of magnification is unknown. The high pollen values for grasses, herbs and heath types combine to suggest a landscape which is still predominantly open, with birch, and a very little pine, established in copses in the more sheltered areas or on the site itself.

The upper part of this zone (from 675 cm) can be interpreted as Godwin's Zone V. As a whole, pollen influx from local sources increases, and long distance transport of pollen is less important (Birks and Birks 1980). The problematic curves for oak and elm in the pre-peat material make it difficult to say if this is the beginning of the Zone, or whether there is an hiatus from the previous samples. Corylus pollen increases suddenly and by a factor of 15, while heathland, grassland and wet habitat features are drastically reduced. This is to some extent a percentage phenomenon, although it is Betula which bears the brunt of the rise in Corylus values, but some reduction in open habitats must have occurred as hazel invaded the hill tops in large numbers. At over 60% TP, hazel pollen represents the dominant species at this altitude, rapidly invading open ground as scrub, or even light woodland, as in the open hazel will form into a small tree rather than a shrub (Godwin 1975a). The high pollen production of hazel in these situations indicates that its dominance of the pollen spectrum is greater than its dominance on the ground. However, the contraction of open habitat species suggests that woodland, presumably with hazel dominant was covering large tracts of ground.

The overrepresentation of birch is corrected, 10% TP giving a truer reflection of its status in the landscape. The reduction from 50% TP may, in fact, mean no reduction in the number of trees in the area (Moore and Bellamy 1974), and may merely indicate that birch is no longer growing on the site itself. Pine values show a slight increase, indicating pine was consolidating its position, but no 'explosion' of its population is recorded, as at many sites to the south and east (Godwin 1975a).

The herbaceous communities diminish rapidly in the Corylus period of SMC 1, both in variety and abundance. Two occurrences of Plantago species, which may be relicts of late glacial short turf communities (Scaife 1982), are the only new indications of continuing late glacial communities. Plantago maritima, however, can also be a natural component of the bank vegetation of high altitude streams (Clapham, Tutin and Warburg 1962). Plantago lanceolata occurs at the same level, and while it is no longer considered to be entirely a cultivation weed (Smith 1970), it is still an unusual component of the early postglacial vegetation (Pigott and Pigott (1963) record it at Malham to the north, during the late glacial itself).

Sphagnum values drop markedly at this time, either due to a reduction in sporing or a reduction in the area of ground it covered. However, both these phenomena would indicate a diminishing number of waterlogged sites in the area (Conway 1954). In one sense, this is difficult to explain, as it is at this point that peat was about to form here on Soyland Moor. However, this site is unusual in that its initial peat formation is caused by very local hydrological conditions (most sites in the area see peat formation much later, during the later part of Zone VIIa (Tallis 1964b)). This zone is thought in general to have been dry and warm (Lamb et al 1966), and this may account for the reduction in Sphagnum.

The latter part of SMC 1, unlike the earlier levels, shows a landscape with little relationship to late glacial vegetation patterns. Hazel dominated the plateau top, with birch and pine interspersed through the scrub or in copses in sheltered area. Grass and heath communities diminished but by no means disappeared. Ground conditions were probably more stable and waterlogged areas that were a result of earlier peri-glacial conditions were drying out to be invaded by dry land plants.

SMC 2 664-636 cm

6700 ± 75 bc (Q 2392) - 6160 ± 50 bc (Q 2391).

Description

Corylus is the dominant species in this zone, swamping the pollen spectrum with up to 80% TP. It never reaches these percentages in the later part of the diagram and the end of this zone marks the end of its overwhelming dominance. Betula maintains the 10-15% TP to which it had declined with the rise of Corylus at the end of SMC 1. Pinus values rise through this zone, overtaking Betula at 640 cm, near the end of the zone, when it attains 12% TP. Ulmus pollen is present in small amounts from the base of the peat profile, and rises to 5% TP during the early part of SMC 2, a position it never consistently surpasses. Quercus does not appear until c. 100 years later, at 6000 bc and its percentages remain low throughout this zone. Sorbus, particularly S. aucuparia first appears in this zone. Salix values, while considerably depleted from the previous zone, are still high in the early part of SMC 2 but begin to decline. Salix pentandra, the bay willow, is also recorded.

The heathland pollen curves show an erratic pattern of peaks and troughs. Calluna is present and twice reaches 10% TP. However, at the beginning of the zone, ericales other than Calluna dominate the heathland species. Empetrum reoccurs, and peaks, later in the zone. The Gramineae and Cyperaceae values are low throughout, rarely reaching beyond 5% TP. Of the herbs, disturbance indicators are common; Cruciferae, Rumex types and Umbelliferae are abundant and occur with Melampyrum, Urtica, Artemisia, Chenopodiaceae and Compositae types. Fresh water and wetland types are seen in cf Ranunculus trichophyllus, Potamogeton, Succisa and Potentilla type. Filipendula is still important and displays an almost continuous curve. Pteridium aquilinum spores are present throughout the zone and

Polypodium vulgare appears at the end.

Interpretation and Correlations

The landscape during 6700-6160 bc was overwhelmingly dominated by hazel. It achieved its maximum at Soyland during this zone, indicating a correlation with the late Zone V or early VIa. Deacon (1974) states that its "explosion" in many areas at this time indicates that hazel must have had refugia in the late glacial period much closer than other trees and shrubs. Southern Europe was often the furthest point north for most temperate species, but hazel may well have remained in small amounts in Ireland, northwest Europe or Britain itself. Interpretation of these hazel values is difficult. Modern analogies with a hazel dominance are rare and its behaviour within differing plant communities is not well-studied. Erdtman (1931) points out its great plasticity of response to differing environmental changes. The most that can be said is that the percentages at Soyland Moor signify that hazel found conditions highly favourable here and took advantage of the open landscape to firmly establish itself.

Birch values, in contrast with early SMC 1, are very low, rarely reaching 10% TP, and pine values become more important by the end of the zone. This is in contrast with a number of highland sites which have much higher Betula values: in Nidderdale, they reach 20-60% TP with pine remaining at 10% TP (Tinsley 1975); Simmons (1969) notes high values in the North York Moors diagrams, and Walker and Lambert (1955) have similar high values at Kirkby Thore in Westmorland. There is some argument as to the real meaning of the birch values. Andersen (1967), Tauber (1967a) and Smith (1972), among others, all state that Betula pollen is over-represented by at least a factor of four. However, it would appear that in certain circumstances birch pollen can actually be under-represented. Tallis (1964a) shows that central and southern Pennine sites have very low Betula values, yet macro remains indicate Betula was a dominant feature of the landscape. Tinsley (1972) confirmed this in modern pollen research. She found that altitude and exposure can significantly reduce birch pollen productivity, and that a 10-15% AP value (not TP) can be indicative of a woodland dominated by birch. In this period, some British sites show high birch values until pine increases, after which birch values decline (eg Bishop Middleham in Durham (Bartley et al 1976)). Turner and Hodgson (1979) statistically tested a large number of sites in the northern Pennines and lowland Durham and found that birch and pine had an inverse relationship which is independent of percentage factors, concluding that the two trees were competing for the same niches, sometimes with birch consistently dominating, and others showing a cyclical relationship. However, this explanation does not fit the facts at Soyland Moor. Birch, after its major 'decline' from the early postglacial shows a remarkably steady pollen curve with no major peaks or troughs. (The slight decrease seen when pine rises dramatically in SMC 3 is most likely attributable to the percentage format of the diagram). Pine cannot be strongly competing with birch, as it achieves its maximum values without disturbing the birch values. In this case, it is possible that birch was occupying the wetter areas of the plateau top, with pine invading into the drier, better-drained rocky outcrops which occur at intervals across the area. The subsequent rises in oak and alder pollen values also do not strongly affect those of birch. Here, particularly in the case of alder, it may be that these more thermophilous elements could not compete strongly enough at this altitude to displace birch. Thus birch, the pioneer tree, maintains a permanent role in the upland woods at Soyland which was never seriously contested by the other trees.

Birch remained the major woodland element in this period, although by the end of the zone pine had established an important local presence. Oak values never reach 5% TP which is likely to indicate it was not yet present on the plateau top (Tinsley and Smith 1974). On the other hand, elm percentages are substantial in this zone, and soon reach 5% TP, which may well indicate its presence in more sheltered areas of the plateau top, or, at the least, the heads of the nearby valleys. Turner and Hodgson (1979) found that elm was often better represented at higher altitudes than in the lowlands, which could contradict the long-held opinion that elm grew only on the more fertile soils (eg Morrison 1959). However, elm species are known to have a complicated ecological behaviour, especially when invading into existing woodland (Rackham 1976), so that the ecological implications of the work of Turner and Hodgson are difficult to judge.

As hazel values reach their maximum willow pollen is reduced to continuous presence only which suggests it was relegated to streamsides or wet hollows. (The occurrence of the bay willow itself confirms this, as it is a common streamside or wetland component today (Clapham, Tutin and Warburg 1962).) Of the dwarf shrubs, it is the ericales rather than *Calluna* itself which dominated the heathland species at the beginning of SMC 1, but are gradually and permanently replaced by *Calluna*. In general, *Calluna* is not an important feature in British pollen diagrams until Zone VIIa or VIIb (Gimingham 1960), even at high altitude sites similar to Soyland. Valley Bog in the north Pennines (Chambers 1978), the North York Moors sites (Simmons 1969) and Rishworth Moor, just 1 km to the southeast of Soyland, all show little or no *Calluna* until just after the Elm Decline. In other areas of the Pennines it is even later, Tallis and Johnson (1980) dating the initial *Calluna* rise in Longdendale, north Derbyshire to 2420-1790 bc. There are one or two sites which show a *Calluna* rise in Zone VI: Malham has a peak of ericaceous pollen in VIc, which is associated with a charcoal layer and an increase in herbs (Pigott and Pigott 1963), and Cranberry Bog in Durham also shows an early peak (Turner and Kershaw 1973). Only Robinson's Moss, Derbyshire (Tallis unpublished) shows a similar rise to that at Soyland. As an oceanic species (Gimingham 1960), conditions in the early postglacial would not have been conducive to the spread of *Calluna*, and it is difficult to explain its unusual early high values at Soyland and Robinson's Moss. It is present in the pre-peat deposits at SMC (and SMD) which indicates early migration into Britain after the glacial maximum. Small populations present in the late glacial could then have increased by invading the edges of the early mires formed at Soyland and Robinson's Moss. From here, early heath communities could have been established in the continuing open areas of the plateau top.

Empetrum reappears and peaks during the latter part of this zone. This is paralleled only at Robinson's Moss, 20 km to the south of Soyland. Usually, *Empetrum* pollen is not represented after Zone IV until after the Elm Decline, although some sites do have consistently low values throughout the postglacial, indicating the existence of small relict communities (Godwin 1975a). However, its reappearance at

Soyland and Robinson's Moss may well be the result of the early establishment of Calluna heath, in which Empetrum could re-establish itself and occasionally become locally dominant, rather than permanent relict communities.

Despite the increasing forest cover, the majority of the herbs confirm the presence of open ground, and also of continuing disturbed areas. Cruciferae types, Rumex species and Umbelliferae are common, and Urtica, Melàmpyrum, Artemisia, Chenopodiaceae and Compositae types are also present. The continuing existence of open water around Soyland is shown by the records of Ranunculus trichophyllus and Potamogeton. Filipendula values are lower but still maintain a continuous profile from the previous zone. Filipendula could be part of the herb layer in damp woodlands, but in these conditions would produce less pollen which would not travel as far (Andersen 1973). Its virtually continuous presence suggests permanent communities on stream banks. The increasing woodland cover is reflected slightly within the composition of the herbs, although, as mentioned, these are less likely to be recorded than open ground species. Pteridium aquilinum values increase a little, suggesting establishment in its natural habitat as a woodland herb (Rymer 1976). From there, it may well have invaded the open heath communities, although its low values indicate that it never became dominant. Polypodium vulgare makes its first, erratic presence. It is a shade-loving fern to be found on the woodland floor, or on the trees as an epiphyte (Mamakowa 1968). The presence of Humulus at 643 cm confirms the presence of moist woodland in the vicinity (Godwin 1975a).

In conclusion, wooded areas were extending, but the woodlands remained open, with possibly one of two sheltered areas developing denser stands of hazel, birch, pine and possibly elm, creating moist and shaded microclimates. Open ground is perforce contracting as shown by the decline in herbaceous pollen values from 50% TP in SMC 1, to 20% TP by the end of this zone. Heath and grassland remained in patches, and the weed flora shows continuing areas of disturbed ground. The date of 6,700 ± 75 bc places the base of SMC 2 in the middle of Godwin's Zone V, and this is reflected in the dominance of hazel and the small but expanding tree values.

SMC 3 636-608 cm

6160 ± 50 bc (Q 2391) - 5690 ± 40 bc (Q 2390)

Description

The beginning of this zone is marked by the rational rise of Quercus pollen (to 10% TP) and by the great expansion of Pinus pollen, which achieves a constant 15-20% TP throughout this period. Correspondingly, Corylus values decline through this zone, from 60% to 30% TP. Betula and Ulmus maintain their values with only a slight percentage decrease. Calluna values increase greatly, while the other ericales virtually disappear. Sphagnum values are low, as is usual in this diagram when Calluna values are high. Gramineae and Cyperaceae values remain low as in the previous zone, although Gramineae begins to rise at the boundary of SMC 3/4. Hedera helix pollen is the major addition to the herbs, establishing an almost continuous curve. Filipendula pollen is still common but declining, whereas Potentilla type pollen, up to now a sporadic contributor to the profile, establishes itself in a continuous curve with high values. It peaks mid-way in the zone at 5% TP and remains high throughout the zone. Succisa and Melampyrum pollen show small peaks. Pteridium values increase and Polypodium vulgare establishes a more continuous record than previously.

Interpretation and Correlations

The major feature of SMC 3 is the expansion of pine. The species involved is Pinus sylvestris (Godwin 1975a), which can thrive in a wide range of habitats from bog to dry sandy soils and in a wide variety of soil fertility (Carlisle and Brown 1968). The history of pine during the postglacial in Britain reflects this range. It is widespread in low values throughout Britain from the late glacial period, although some sites in both northern and southern Britain show early increases, leading Godwin (1975b) to postulate two discrete centres of spread. However, the pine rise at Soyland is dated to 6160 ± 50 bc, correlating with Zone VI, the 'standard' period of pine expansion (Godwin 1975a). Robinson's Moss (Tallis unpublished) shows a similar pattern, with a date of 5725 ± 65 bc for the period just after the initial pine rise. Carlisle and Brown (1968) state that pine often develops in pure stands, with only holly and willow interspersed throughout. At the beginning of the pine maximum at Soyland, it is interesting to note a small peak of willow pollen, and that two out of the three occurrences of holly pollen in the profile are also positioned here. These two species may well have spread into the pine stands, and their loss as pine increases may be more apparent than real: both species are insect-pollinated, with low pollen production and dispersal, and, as the pine stands developed and spread, pollen dispersal of other species would remain within the woodland and not fall on the bog (Andersen 1967).

Oak values increase through the zone, perhaps at a slower rate than other sites. Andersen (1967, 1973) states that oak pollen is over-represented by a factor of four, and this figure has become common credence, despite the many studies which argue that Quercus pollen profiles match the position of oak in the landscape, (eg Smith 1972, Hansen 1949), and may even be under-represented (Bradshaw 1981). Taking the oak curve at its face value the 10% TP suggests oak had reached the plateau top itself if not yet in any great numbers (Tinsley 1972). Certainly, it had not yet invaded into the pine stands dominating the local woodland. The slight decline in birch and elm pollen at this point is likely to be a statistical drop due to the increase in the pine pollen, so these two species in reality maintained their respective positions in the landscape.

The contraction in hazel scrub seen here is a little earlier than many English sites (Godwin 1975a). The major cause may be the expansion of pine. Hazel may have been an understorey shrub in the pine stands, in which case its decline is more apparent than real, partly due to its reduced flowering capacity in shade (Godwin 1975b), and partly because its pollen is less likely to reach the bog surface (Andersen 1967). However, the decrease in hazel values must represent a genuine decline in numbers. Favourable areas for hazel were being invaded by larger trees, and, as an understorey shrub, hazel would be less vigorous and less able to regenerate (Edlin 1970, Dimbleby 1978). Despite its reduction by half, at 30% TP hazel was still important, and indeed was still the dominant species in the area. Thus hazel may be occupying two habitats on Soyland during this zone. Firstly in its pioneer role as open scrub or woodland on the plateau top, and secondly, as an understorey shrub among the increased area of oak and pine woodlands.

There is one occurrence of Prunus type here. Prunus spinosa, the blackthorn, is the most likely species to have been incorporated, as all Mesolithic macro records of Prunus are sloe fruit stones and blackthorn charcoal (Dimbleby 1978, Godwin 1975a). The three isolated occurrences of Prunus type through SMC may indicate a reasonably continuous presence, as its pollen production is sparse (Godwin 1975a). Blackthorn may have been part of the hazel community, and may have invaded into the edges of the increasing woodland.

The herb component reflects the increased areas of woodland. Hedera helix appears for the first time and quickly establishes itself. It is a poor pollen producer and disperser (Troels-Smith 1960), and its virtually continuous presence in the profile suggests it was common. Although it is commonly seen as a species growing in the primary woodland (eg Iversen 1944), it is far more common in either secondary or open woods, or on the edges of woodland (Rackham 1976). Bracken also increases, and Polypodium vulgare shows a more constant presence. Aquatic and wetland herbs are seen in Typha, cf Ranunculus trichophyllus, Lysimachia vulgaris type (which occurs for the first time), Filipendula and lastly Geum. The latter probably represents Geum rivale (Godwin 1975a) and is a herb of streamsides or moist woodlands. Indicators of disturbed ground are still important and include one grain of Carduus. This is usually found in the late glacial or post Elm Decline period, and is another disturbed ground indicator which found the higher altitudes a suitable refugium during the wooded period of the early and middle postglacial (Godwin 1949). The three peaks of Succisa, Melampyrum and especially Potentilla can be explained as representing a clearance phase and as such will be discussed later.

SMC 3 sees the beginning of the woodland expansion on the high ground with pine as the dominant species. The actual area of ground with tree cover, and the role of hazel and heathland is far more difficult to gauge than in the previous period, but on the whole open or lightly wooded ground was still the probable dominant feature of the landscape.

SMC 4 609-541 cm

5690 ± 40 bc (Q 2390) - 4160 ± 40 bc (Q 2387)

This zone is sub-divided at 583 cm.

SMC 4a 609-583 cm

5690 ± 40 bc (Q 2390) - 5025 ± 40 bc (Q 2389)

Description

The beginning of this zone is marked by the rational rise of Alnus pollen (as defined by Smith and Pilcher 1973). It rises gradually through the sub-zone reaching 12% TP at the boundary with SMC 4b. Pinus pollen is high throughout at 20% TP. Tilia first appears here, but its presence is not yet constant and only single grains are recorded. Fraxinus appears for the first time also, even more sporadically than Tilia. Betula, Quercus and Ulmus maintain their SMC 3 values. Corylus continues to decline at the beginning of this zone, then partially recovers, regaining 40% TP. Salix declines slightly to its now permanent position of continuous presence only. Calluna shows a much less erratic curve than previously, maintaining 10-15% TP throughout. Cyperaceae remains low, but Gramineae peaks early, reaching 20% TP before declining to its more usual position of less than 2% TP.

Of the herbs Potentilla type values are still high from the peak in SMC 3, but decline to continuous presence only by the SMC 4a/b boundary. The full range of open and disturbed ground herbs is present, but only the Compositae types show any abundance. One grain of Plantago lanceolata occurs at 587 cm. Hedera helix is present but not as often as in SMC 3.

Interpretation and Correlations

Arboreal pollen values, excluding hazel, reach 47% TP by the end of SMC 4a. This figure is only occasionally improved upon, implying that, from this point until the Elm Decline, woodland at this altitude changed only in its relative components, not in terms of actual ground cover.

The beginning of this sub-zone marks the rational rise of alder values at Soyland (just after 5690 ± 40 bc). In most British pollen diagrams this feature characterises the BAT (the Boreal-Atlantic transition or the VI/VIIa boundary). Here, however, the classic sequence of the BAT, of an abrupt alder rise with a concomitant pine decline, is not seen. At Soyland, alder rises slowly and gradually and pine remains at its maximum values, not declining until much later at 4160 ± 40 bc. The dates for the BAT in Britain fall between 5600 bc and 5200 bc (Smith and Pilcher 1973). At Soyland, alder rises in the very early part of the sequence, which is more usual at sites without high pine values (Oldfield 1965). The pine decline at Soyland appears to be unconnected to the BAT.

One of the major problems in reconstructing the position of pine in the landscape lies in the local nature of its pollen representation. Once thought to be a major long distance transport pollen type, it is now agreed that pine pollen travels no great distance from its source (Godwin 1975a). Sites very close together can show substantially different profiles (eg. Kildale Hall Sites A and B, in Cleveland (Jones 1977) and Chat Moss B and C, in Lancashire (Birks 1964)). Hodgson (1974) points to a range of 6%-90% AP for pine during the Boreal in the north Pennines, and Turner et al (1973) state that it is the only major variable during the Boreal in Upper Teesdale. This variability makes general statements on the status of pine over wide areas very suspect. Thus, in the northern Pennines and lowland Durham, Hodgson (1974) relates high pine values to higher altitudes, whereas Turner and Hodgson (1979) state the opposite. It is therefore possible that the extended pine maximum at Soyland (at both SMC and SMD) is a relatively local phenomenon only. It is not likely to represent pine growing on, or on the immediate edges of the bog, as in these circumstances the pine profile would be highly erratic (Janssen 1967) and the values much higher (O'Sullivan 1973), at least in relation to the rest of the arboreal pollen. Thus, the long pine maximum represents pine continuing to grow in large stands across the plateau top, in the vicinity of the bog. As discussed previously, the lack of competition with birch suggests pine may be occupying the well-drained outcrops which occur around the bog.

When looking at other sites in the central and southern Pennines, the local nature of the pine profile is underlined, although some generalisation is possible. The majority of the pollen sites in the area start at (or after) the BAT and, where it is recorded, show a standard and contemporaneous alder rise and pine decline (eg Totley Moss, Derbyshire (Hicks 1971)). Sites where peat accumulation begins earlier are tending to show an extended pine maximum. Robinson's Moss in Longdendale (Tallis unpublished) shows an identical spectrum to that of Soyland, coupled with the relatively early, gradual alder rise. Bradwell Sitch Site A, only 4 km south of Robinson's Moss (Tallis and Johnson 1980) has high pine values until the post-Elm Decline period, and with a much later although still gradual alder rise. In the former set of sites, these differences may be partly explained by the low resolution of the pollen data due to slow rates of peat growth, the lack of radiocarbon dates, and the lack of information from the preceding period, which may combine to make what is in reality a gradual change appear abrupt. However, this is not always the case, and one site, Fountain's Earth in Nidderdale (Tinsley 1975), shows the standard sequence despite its early date for the onset of peat accumulation.

It is possible that the base-poor soils of the millstone grit, which may have been rendered more acid by the pine stands themselves (Taylor 1965, but refuted by Dimbleby 1962a), were a relevant factor in the extension of the pine domination. Both at Robinson's Moss and Soyland, oak values remain constant during the pine maximum, only expanding to its highest values after pine has declined. This is unusual, as most sites show the maximum values for oak immediately after the BAT. The values for oak at Robinson's Moss and Soyland indicate it was present around the sites during the pine dominance, but took much longer to displace pine from its circumscribed areas. Acidic soils, while not preventing oak from invading, could considerably delay its expansion when in competition with pine. However, as an explanation, this theory is tenable only for this area, as Turner <u>et al</u> (1973) and Bartley <u>et al</u> (1976) correlate high pine values with base-rich soils in the northern Pennines and Co. Durham.

To summarise, pine values can only be interpreted for the area around Soyland, where it remained the dominant tree species for a much longer period than at most British sites, and delayed the expansion of the small oak communities established on the plateau top. Evidence from Robinson's Moss may indicate that these features were relatively widespread in the central Pennines, but more data is necessary to confirm this.

The empirical limit for alder occurs in the previous zone and its rational limit occurs just after 5690 ± 40 bc. This date is compatible, if on the early side, with other British sites (Smith and Pilcher 1973). However, most sites with early alder rises, such as Mordon Carr in Durham with a date of 5809 ± 67 bc (Bartley <u>et al</u> 1976), usually do not see an increase in pine at any point in the profile. Conversely, those sites with an extended pine maximum tend to show a date for the alder rise nearer to the Elm Decline (Oldfield 1965). (Turner and Hodgson (1979) state that these phenomena are genuine and not statistically produced.) This would seem to suggest an inverse relationship between pine and alder populations which is not ecologically very likely, and is certainly not seen at Soyland Moor (or Robinson's Moss), where pine maintains its dominance as alder gradually establishes itself.

A closer look at the spread of dates across Britain shows much variability, with altitude, latitude and soils playing no definitive role. Thus alder was common in the Scottish highlands prior to the standard BAT (Godwin 1975a), and Beckett (1981) shows both early and late dates for the rational limit of alder in lowland Humberside. Thus, as with pine, the initial establishment, mode of expansion and quantity of alder was highly variable both spatially and temporally. McVean (1956) points out that the establishment of alder seedlings is controlled by a large number of factors which could explain the differing responses. At Soyland Moor, alder's relatively early rise could be explained by its establishment and spread up the deep and sheltered valleys around Soyland. The lack of a sudden rise as at many other sites could be due to the late frosts and strong winds which would affect the relatively open plateau area, and can inhibit alder seedling establishment (McVean 1956).

Lime and ash appear in the profile for the first time. Even allowing for their poor pollen productivity and dispersal (Grieg 1982, Andersen 1973), the tiny quantities suggest that neither species was present on the plateau, although there may well have been a small number of these trees in the valleys.

The continuing decline of hazel again may be partly due to the increase in trees. However, when hazel values rise at the end of this sub-zone, there is no accompanying decline in the trees. The herb and dwarf shrub profile suggests an increased area of open ground, despite the slight extension of the woodland and thus heathland and rough grass may be extending at the expense of hazel scrub. There is a full complement of ruderals, weeds and open ground herbs. Conversely, the herbs do not indicate the increasing role of woodland. Only the sporadic occurrence of ivy, and the now continuous presence of Polypodium vulgare represent the woodland ground flora. Bracken does increase in this zone but whether as a woodland herb, or as an invader into open areas, is difficult to say.

In this sub-zone we see woodland values reach their maximum constant level. However, these values indicate that woodland remained open. The contraction of hazel scrub also allowed an increased area of grass and heathland, leaving a more highly contrasting landscape. The role of man in this pattern will be discussed in a later chapter. The woodland itself was still dominated by pine, keeping oak in a secondary role. Alder arrived in the area, establishing itself slowly up the valleys along the water course banks, and into the woodland.

SMC 4b 583-541 cm

5025 ± 40 bc (Q 2389) - 4160 ± 40 bc (Q 2387)

Description

The major change seen here is the end of the pine maximum. By the end of this sub-zone its values have declined to 5% TP. *Betula* values also show a very slight decline. Both *Quercus* and *Alnus* values exhibit a very gradual rise, the latter just overtaking *Quercus* by the end (at 17% TP and 15% TP respectively). *Tilia* values too show a gradual increase and the *Fraxinus* curve becomes continuous. One grain of *Acer*, its only appearance in this profile, occurs at 548 cm. Two grains of *Populus* also occur.

Corylus values have reached their second, though lesser, maximum, with an average 35% TP throughout the zone, although the curve is erratic, showing minor peaks and troughs. The ericaceous curves show a marked decline. *Calluna* values themselves decline to their lowest point, at under 2% TP for most of the sub-zone. Towards the end they increase again, reaching 20% TP at the SMC 4/5 boundary. Both Gramineae and Cyperaceae show erratic curves, although they are generally higher than previously.

Potentilla type shows a second peak at the end of the zone dated to 4390 ± 50 bc (Q 2388). It has been continuously present since its first peak, and rises again to reach 9% TP at 548 cm. Some of the features associated with the first peak also reappear. A small *Succisa* peak occurs, as before, just below the *Potentilla* rise and *Melampyrum* is present in three contiguous levels, coincident with *Potentilla*. A new feature here is the increase in the occurrence of Chenopodiaceae, and one grain of *Plantago lanceolata* occurs at 549 cm. *Artemisia* is frequent, but occurs below the *Potentilla* rise, and other disturbance indicators are scattered through the sub-zone. Wetland plants become common: *Typha/Sparganium* pollen is abundant and *Caltha*, *Ranunculus* cf *trichophyllus* and *Nuphar* occur. Possible streamside species are present in *Chrysoplenium*, *Lysimachia vulgaris* type and *Filipendula*. *Drosera rotundifolia* represents part of the mire surface vegetation. *Hedera helix* is common and *Humulus* occurs three times during this zone. The *Polypodium vulgare* curve shows a slight increase, but *Pteridium* is declining and shows an erratic curve.

Interpretation and Correlations

For the major part of this sub-zone, the woodland values remain at the maximum seen at the end of SMC 4a (40-50% TP). They decline to 32% TP during the upper part, coincident with the *Potentilla* peak. This is dated to 4390 ± bc, and may represent a clearance phase. Pine remained important but declined as alder and oak began to dominate the woodland. Birch also declined slightly. The alder values suggest it is unlikely that alder carr was present on or close to the edge of the Soyland peat deposits, as in these conditions *Alnus* pollen would swamp the diagram (Janssen 1959). Cromwell Bottom, Elland, about 12 km north east of Soyland, shows alder at 83% AP and clearly alder carr was developing on or about the site (Bartley 1964). The most likely habitat for alder here was on the streamsides and in wet depressions, where it may have competed with birch. The decline in willow as alder increased suggests that this might be the case, alder shading out or replacing some of the willow stands as it reached up the valleys onto the plateau top. The low values of willow are due to its poor pollen production and the continuous presence of single grains indicates it was still likely to be growing nearby the peat deposits (Bradshaw 1981).

There is evidence for increasing wetness during this period: the increase in alder itself, the large rise in *Sphagnum* and the increasing range and occurrence of wetland herbs, and those liking moist conditions, such as *Typha/Sparganium* and *Caltha*. The date for the onset of this wet period is 5025 ± 40 bc. This change has often been associated with the BAT, the rise in alder explained as an increase in favourable conditions for its expansion, and the decline in pine seen also as a result of increasing wetness. However, this is not the case at Soyland Moor, as discussed previously, where these two changes take over 800 years to be accomplished. The sustained high values of alder at this time may well be associated with this increased wetness, but its initial rise is unconnected. It is possible that surface moisture in the deep sheltered valleys which dissect the plateau was adequate to allow alder to establish there as soon as migration rates allow, and to let it slowly increase onto the plateau from a well-established base. Thus the added wetness in this period merely boosted its competitive status and allowed it to continue to increase in numbers and to invade at higher altitudes.

The decline in pine may not be related to the increasingly wet conditions, as it is a species which can tolerate extremes of soil moisture (Carlisle and Brown 1968). Also, if pine was occupying the better drained areas of the plateau, the increasing wetness would not have had as great an effect. This is supported by Oldfield (1967) who points out that pine had two diametrically opposed responses to an increase in wetness: expansion and decline. At Soyland Moor, it is more likely that pine was being replaced by the increasing amount of oak, as pine is a species susceptible to competition from deciduous trees (Godwin 1975a). The slow spread of oak and the extended pine maximum may suggest that conditions were at the limit for the expansion of deciduous woodland (Smith 1965). The arrival of *Acer*, which is early in comparison to most sites, the post-Elm Decline period being more usual, may in fact contradict this. However, *Tilia* shows a relatively late arrival, both empirically and rationally. Its rational limit at 4390 ± 50 bc is in general 400-500 years later than other sites. For example Bishop Middleham, Durham has a date of 4810 ± 120 bc (Bartley *et al* 1976). Although this is a lowland site, it is situated much further to the north. Nant Ffrancon in Wales has a date of 4775 ± 100 bc (Hibbert and Switsur 1976). This delay in arrival at Soyland might suggest difficulties in initial establishment and spread for species demanding warmer or less exposed conditions than pine or birch. The occurrence of two grains of *Populus* here (as well as the large amounts of hazel) suggests woodland was still very open. The species is likely to be *Populus tremens*, which is light demanding (Godwin 1975a). Its absence through most of the diagrams is misleading as the pollen is highly degradable (Litchi-Federovich and Ritchie 1965), and it is likely to have been present here throughout the postglacial.

The increase in hazel may be significant in relation to the decline of pine, re-occupying some of the areas inhabited by pine. Smith (1965) suggests that hazel, once established, exerts a considerable inertia, and even when reduced by competition, will expand again at the first opportunity. This would appear to be a highland phenomenon; sites from lower ground show that hazel was important only in its initial seral role and declined permanently as woodland species invade, eg the lowland Durham sites (Bartley *et al*

1976). However, it is also possible that hazel was merely re-occupying those areas of heathland which extended in the previous zone. The temporary decline in hazel at the end of this zone is correlated with an increase in heather, confirming their relationship. This latter phase represents a probable clearance episode, and may suggest that hazel was not an important factor in the decline of pine.

This sub-zone saw the extension of deciduous forest in the uplands, although as yet it was still not dominant. Heathland and grassland declined as hazel reestablished itself.

Zone SMC 4 as a whole is a period which has very few correlations with other sites, apart from the nearby Robinson's Moss. It represents a long transition phase between Zone VI, a period of seral woodland, and Zone VIIa, the full expression of deciduous forest in Britain.

SMC 5 541-469 cm

4160 ± 40 bc (Q 2387) - 2615 ± 50 bc (Q 2385)

Description

Betula values remain at the slightly lower level of the previous zone, while Pinus values quickly decline to their lowest point in the profile, never reaching 5% TP. Quercus values continue their steady increase, reaching 25% TP midway through this zone. At 501 cm these decrease sharply to 15% TP. Although they recover immediately, they never again reach the high values of the early part of the zone. Alnus pollen reaches its maximum values at the beginning (15-20% TP). They too show a regression at 501 cm, but fully recover immediately. However, this is not maintained and for the latter part of the zone alder values remain at 10-15% TP. Tilia pollen reaches its maximum of 3% TP towards the middle of the zone, dated to 2915 ± 50 bc (Q 2386); these values are maintained. Fraxinus values rise too, reaching 2% TP. Its curve is not as smooth as Tilia's, showing two peaks, one at c.510 cm and a longer, more gradual peak around 495 cm, after which it declines slightly.

Of the shrubs, Crataegus (see Appendix III for identification), Prunus and Ulex types are recorded in this zone. Sorbus species, although not as common as previously, are still present. The Corylus profile shows an erratic pattern, peaking in the initial stages after its decline, in the middle and at the end of the zone. At these three points it reaches 30% TP. In the intervening troughs it declines to 15% TP, its lowest point since its first migration into the area. The Calluna profile shows a rough negative correlation with that of Corylus until 500 cm, after which it maintains a high value of 20% TP, showing only a slight drop when Corylus values rise at the end of the zone. Empetrum and ericales values show peaks at the same level as the final Calluna rise, but they are not maintained beyond 490 cm.

Gramineae and Cyperaceae have greatly differing profiles for the first time. The Gramineae profile shows an increase in the first part of the zone, occasionally reaching its maximum of 25% TP, but then drops to a very low value of 3%. The Cyperaceae profile is more erratic, but generally shows low values for the first half of the zone. It peaks at the same time, and roughly for the same duration, as Empetrum and the ericales. It declines to less that 5% TP by the end of the zone.

The herbs show a greater diversity and quantity than in immediately previous zones. Viscum album and Menyanthes trifoliata appear for the first time. Typha/Sparganium shows a peak in this zone. Pteridium shows a much more erratic profile than previously, with two peaks evident. In the earlier peak, midway through the zone, it reaches 14% TP. Conversely, in between the peaks, it shows lower values than in previous zones, interspersed with occasional high values. Artemisia, Chenopodiaceae and Compositae types are common, which Rumex acetosa/acetosella shows an almost continuous curve after 509 cm. Plantago lanceolata occurs often after 499 cm and P. major/media is also recorded. The rare Mercurialis cf annua (see Appendix III) occurs here. Potentilla type again produces an almost continuous profile, including one small peak at the beginning of the zone (530-540 cm).

Interpretation and Correlations

The early part of this zone (4160 ± 40 bc -2915 ± 50 bc) shows the maximum values of deciduous forest in the area around Soyland Moor, with oak and alder dominating the profile. Oak only reached its full extent between 3500-3000 bc which is late in comparison to most British sites, where the full extent of oak is dated to 5000-4000 bc. As discussed, the decisive factor would appear to be its early inability to compete with pine which was well-established in the vicinity. Where pine was low throughout, eg Neasham Fen, Durham (Bartley et al 1976), or where oak was well-established prior to the invasion of pine, eg Nant Ffrancon, Gwyned (Hibbert and Switsur 1976), oak reached its maximum extent earlier. A similar situation to Soyland is found at Tregaron Bog, Dyffed (Hibbert and Switsur 1976), where oak reached its maximum at 3160 ± 70 bc. Interestingly, at Tregaron, oak managed to temporarily displace pine at an earlier period, but pine regained its dominance. Tregaron is at 164 metres OD, midway between lowland and highland. These examples suggest that pine is in direct competition with oak, and that higher altitudes were more favourable for pine initially. The mid-way position of Tregaron could then explain the temporary displacement of pine. At this period on Soyland, oak managed to displace pine entirely on the plateau, the low values of pine suggesting it no longer grew locally.

Alder, too, while never as important as oak, must have been a major presence along the numerous watercourses flowing from the plateau top and in damp or waterlogged depressions. Elm and birch maintained their position. The later woodland invaders of ash and lime were still expanding in the initial part of the zone. The lime species was likely to be Tilia cordata which will occur further north, and on a wider range of soils than T. platyphyllos (Mittre 1971). Lime shows quite a high percentage for this altitude. Converted to arboreal pollen percentages, it reaches 5-7%, which is usual for more southerly, lower sites (Godwin 1975a). Rishworth Moor close by Soyland, shows values of less than 5% AP (Bartley 1975), and Robinson's Moss to the south only once reached 5% AP (Tallis unpublished). Although a little lower than at Soyland, these Tilia percentages are still regarded as high for both the latitude and altitude, as to the north the Tilia record appears to become discontinuous (Moore 1977a): Tarn Moss at Malham shows only traces of lime and never a continuous curve (Pigott and Pigott 1963). Fountain's Earth in Nidderdale shows a continuous curve, but the values are very low (Tinsley 1975). However, parts of the North York Moors (eg Egton High Moor (Simmons 1969)) and some areas of lowland Co. Durham (Bartley 1966) show high _Tilia_ values, sometimes as high as 30% AP.

The amount of pollen produced by lime is commonly seen as being very low, as an insect-pollinated species (Godwin 1975a). Others suggest that it is in fact as high a producer as oak (Hyde and Williams 1945). However, the major limiting factor is its extremely poor pollen dispersal. Flowering does not occur until the leaves are fully open, unlike most trees, and the pollen itself has a propensity to clump, making it heavy (Grieg 1982). Andersen (1973) suggests that, to counter this, the lime pollen values should be doubled to give its true representation, while others suggest a multiple as high as eight (Grieg 1982). Whichever multiple is used, it is clear that a _Tilia_ profile is representative of lime trees close to the site, within a few hundred metres at most (Pigott and Huntley 1980). The disjunct pattern of high _Tilia_ values in northern Britain reflects this local representation. Lime must have been quite an important element in the woodland of the nearby deep valleys around Soyland Moor. These could provide shelter from exposure, perhaps allowing lime to extend to slightly higher altitudes than in more undulating landscapes. From here, it may have established small copses in sheltered areas on the plateau itself, or interspersed within denser areas of oak woodland.

Ash is most likely to have been part of the open woodland on the hill slopes and tops, as it is a light-demanding tree (Godwin 1975a). Its more erratic curve suggests it responded to changes in the successional stages of woodland clearings (whether anthropogenic or natural). The one record of hawthorn and the two of _Ulex_ type confirm the presence of open ground within the woodland. _Ulex_ type includes gorse, broom and _Genista_ species, all of which show preference for open heath habitats today. Today and historically, both _Crataegus_ and _Ulex_ type species are found in connection with human activities, but they are also found in natural clearings.

The latter half of this zone shows a definite reduction in the area of woodland as a whole, and oak and alder in particular. This is associated with a rise in ericaceous acid heath, and a reduction in hazel. These changes can be interpreted as a result of the very short clearance phase at 500 cm (dated to 2915 ± 50 bc), and as such will be discussed in detail later. Hazel values show a general decline through this zone, albeit in an erratic fashion. Pollen diagrams for this period show no standard pattern for hazel, indicating it responded sharply to local factors such as the extent of woodland and heath and the activities of man in the area.

The peak of ericaceous pollen after 500 cm indicates a short period of acid heathland developing after the clearance. As _Calluna_ pollen is grossly over-represented in comparison with the other ericales (Bradshaw 1981, Tinsley and Smith 1974), the high percentage of other ericales here suggest their presence was widespread and dominant within the heathland. _Vaccinium_ was likely to have been the major contributor (Godwin 1975a). After the period of dominance, _Calluna_ values continue to be high and indicate a more permanent extension of heather moorland. This is common to a number of sites during Zone VIIa.

The greater diversity and quantity of the herbs in this upper part of SMC 5 strengthens the hypothesis of openings within the woodland. The full range of disturbance and open indicators is present, such as Plantago lanceolata, and includes some of the more unusual species, such as Rumex obtusifolius type. The erratic nature of the Pteridium curve suggests bracken responded to small successive openings in the woodland, either by extending into open ground, or by producing more spores, with a greater potential for their dispersal. The more open nature of the immediate vicinity may have allowed herbs from other habitats to be trapped in the peat deposits, although, of course, they may merely represent a genuine increase in these species. Thus, the records for ivy are more common than before, and mistletoe is recorded for the first (and only) time. This latter is indicative of high summer temperatures (Iversen 1944) and confirms the evidence of Tilia. The bog surface vegetation is well-recorded and includes Menyanthes trifoliata. The sedge type of Rhynchospora which is likely to refer to R. alba alone (Godwin 1975a), occurs at the end of this zone. This species is usually found in pools on acid mires.

In vegetational terms, this zone compares with Godwin's Zone VIIa, the period when deciduous forest and warmth-demanding species reached their maximum extent. This period begins very late at Soyland (4160 ±40 bc) compared to most British sites (5500-5000 bc). However, the evidence for increasing wetness associated with Godwin's Zone VIIa occurs earlier, in SMC 4b. This implies that increasing wetness was not an overriding factor in deciduous woodland development at Soyland.

Soyland Moor Site D

This is the fourth and final core in the series taken 100 m from SMC. The peat measured 664 cm in depth. Together SMC and SMD straddle the deepest part of the basin. Peat sampling was problematic at this site as much of the peat below the surface was too wet to form a core. The pollen diagram for SMD is from two series of cores taken 2 m apart, and married at 595 cm. Possible problems of matching are discussed in the interpretation of SMD 3. The pre-peat material was hard packed, allowing only 10 cm of mineral soil to be sampled. However, the firmness of the material allowed sampling at closer intervals, at every centimetre as in the peat profile. As in SMC, the peat was amorphous in character until 550 cm, above which it comprised Eriophorum vaginatum alternating with Sphagnum. As discussed in the introduction to this chapter, the pre-peat material will be discussed as an entity, whereas the peat samples will be compared with Soyland Moor Site C only. Highly local differences, however, will be discussed in detail in the next chapter.

SMD 1 677-667 cm Pre-peat material

Clear evidence for contamination is rare in this pre-peat zone. The consistent occurrence of Quercus with no Ulmus may be contamination. Pteridium has a higher curve than in the next zone, and Polypodium vulgare is present throughout. Otherwise the zone shows a consistent profile.

Description

Betula is a major feature of SMD 1; rising from 15% TP at the base to 40-50% TP at the end of the zone. Pinus pollen is consistently present, although under 5% TP. Ulmus is absent, but Quercus is present in very small quantities. Alnus appears in the lowest four levels only. Corylus pollen is present at under 5% TP, and Salix values rise to nearly 4% near the end of the zone, but immediately begin to decline. All the ericales show low values, contrasting with Gramineae values which contribute over 40% TP at the base, but decline in the upper part, ending the zone at 11% TP. The Cyperaceae profile is slightly erratic, averaging 10% TP, until the end of the zone where it increases to 20%.

The major contributors to the herbs are Filipendula (up to 3% TP), Potentilla types (up to 10% TP) and Rumex acetosa/acetosella (up to15% TP). Other herbs which occur frequently are Compositae types, (including Artemisia), Umbelliferae, Ranunculus acris type and Valeriana officinalis. Cruciferae, Chenopodiaceae, Trifolium and Lotus appear sporadically. Thalictrum, Pinguicula, Saxifraga oppositifolia type, Chrysosplenium, Hypericum perforatum and Polemonium caeruleum are types which occur in the later part of the zone. There are also a number of rare species recorded here (see Appendix III for identification): cf Polygonum viviparum, cf Sedum rosea, cf Oxyria digyna, cf Artemisia norvegica, cf Dianthus and cf Illecebrum verticillatum.

Of the pteridophytes, Lycopodium alpinum and Osmunda regalis occur. Polypodiaceae presents a high continuous spore curve of up to 20% TP. Pteridium is present and Polypodium vulgare appears sporadically. Sphagnum overshadows the whole profile. Its percentages were calculated outside the pollen sum because of its erratic sporing, and to stop it swamping the diagrams. In the basal sample, Sphagnum reaches 1240%. It declines through the zone, ending it at a mere 140%.

Interpretation and Correlations

The levels 667-671 cm may possibly be dated to pre-8,300 bc, the end of the late glacial, changing at 671 cm to the postglacial period. The lower birch, higher grasses and lower Polypodiaceae suggest this early, late glacial date, but the occurrence of Quercus and the few grains of Alnus militate against this. The continuous presence of Corylus throughout also suggests a post-8300 bc date. However, Godwin (1975a) points to a number of late glacial sites, especially in the north, where 2-5%TP of Corylus is common. As the changes through this zone are slight, indicating only an increase in birch woodland with open grassland contracting, this zone is treated as one period, dated between 8500-7500 bc, spanning the very end of the late glacial and the beginning of the postglacial. The problems of possible contamination and mixing of deposits do not allow a more definitive correlation to be made.

The lack of major changes, especially in the herb spectrum, indicates that the immediate postglacial period at Soyland retains most of the late glacial features. The increasing birch values may reflect long distance transport from the lowlands (Godwin 1975a), but could also indicate birch growing on the site itself. The significant willow values may represent willow carr on the site itself also, as the pollen type included no grains of the *Salix herbacea* type, which would have represented bushes growing on open grassland (Pennington 1969). This is further discussed in the next chapter. The record of cf *Genista* in the basal sample indicates some open heath (Clapham, Tutin and Warburg 1962), as does the presence of ericales, but this was not extensive.

The continuing existence of late snow beds is implied by *Polygonum viviparum, Lycopodium alpinum* and cf *Oxyria digyna.* The high values of *Rumex acetosa/acetosella* consist mainly of *R. acetosa* (det. by P D Moore), and as such suggest extensive tall herb communities. The inability to distinguish species within *Potentilla* type makes it impossible to state the position it holds in the landscape, but its high values are common in the late glacial and Zone IV (Godwin 1975a). Many of the species which occur sporadically in this zone are more usually found in the late glacial, although their occurrence in Zone IV is not uncommon. Cf *Dianthus* is recorded only in the mid and late glacial (Godwin 1975a). *Polemonium caeruleum,* although only sporadically represented in pollen spectra due to its poor production and dispersal, was a common constituent of late glacial communities. Some species are present which only survive after the late glacial in mountainous areas: cf *Sedum rosea,* cf *Artemisia norvegica, Saxifraga oppositifolia* type and *Chrysosplenium. Chrysosplenium* has very poor dispersal, and its presence in the immediate vicinity is indicated (Godwin 1975a).

The occurrence of *Pinguicula* is rare (Godwin (1975a) gives no records for it). Here it is probably representative of the wet ground around local watercourses which is inhabited by the short turf communities. The two grains of *Centaurium* are also rare, only recorded at Old Buckenham Mere, Norfolk and the Shetlands (Godwin 1975a). The record of cf *Illecebrum verticillatum* (det. by P D Moore) would appear to be unique. It is a self-pollinating rare plant, confined today to a few counties in southern England (Clapham, Tutin and Warburg 1962). It is indicative of sandy, disturbed soils and as such would find favourable habitats in the Soyland area at this period. Most of the above-mentioned herbs are rare today or confined to mountainous, arctic-alpine habitats. They are mainly characterised by their intolerance to competition, and indicate that vegetation cover is sparse. The wide range of disturbed ground indicators suggests soil conditions are still unstable.

This zone is very similar to Zones III and IV (CMCP 3 and 4) at Crose Mere in Shropshire (Beales 1980). Pow Hill in lowland Durham shows a similar sequence, but has a high *Empetrum* record (Turner and Hodgson 1981), and Weelhead Moss II in Upper Teesdale differs only in its high values for *Pinus* (Turner *et al* 1973). SMD 1 perhaps shows a wider range of herb types, suggesting that it was possibly more open. This is not indicated by the AP percentages, but if long distance transport is an important factor then these percentages are not a good guide. A more open landscape would allow herb pollen from a larger area to be incorporated in the soil. It is also possible, given that this area was not fully glaciated for most of the Weichselian, that a more diverse flora had time to develop than in areas recently covered in ice.

Correlations with SMC 1

The upper part of SMC 1, with high Corylus and low Betula, is entirely missing from SMD 1. It is probable that all of SMD 1 is older than even the lower part of SMC 1. The inversion of the Betula curves, the low Pinus, the lack of Ulmus and the continuously high Sphagnum indicates this. The herb spectrum for SMC 1 confirms the earlier date for SMD 1. While the majority of species are similar in range and quantity (eg Filipendula), the SMD spectrum shows a much longer species list, some of which are true arctic-alpine species and most of which are rare today. Most of the species of SMC 1 herb spectrum are open or disturbed ground indicators which became common again after widespread forest clearance, as they were merely limited in the early postglacial by the shade of the spreading woodland. Many of the SMD 1 herbs are limited by temperature and intolerance to competition from vigorous herbaceous growth, and as such are cold-period indicators. The very high value of Potentilla type and Rumex acetosa/acetosella are not seen in SMC 1. This could reflect a local development around SMD, but is more consistent with the earlier date.

SMD 2 664-598.5 cm

Date for base of peat at 664 cm: 6240 ± 70 bc (Q 2395)

Correlations with SMC 2

The pollen evidence for the onset of peat accumulation at SMD suggests a date of between 7600-7000 bc, some time before peat accumulation at SMC. The radiocarbon date is later, but this may be due to the difficulties of dating the base of a peat profile (V R Switsur, pers comm). The radiocarbon date for the base of SMC 1 accords well with the pollen evidence.

The evidence for an earlier date at SMD than suggested by the radiocarbon date is comprehensive. The earlier pre-peat profile points that way, and the very high birch values at the beginning of SMD 2 are reminiscent of Zone IV rather than the Zone V of SMC 1, as are the low pine values. The rational limit of Ulmus occurs well into this zone and only reaches percentages comparable to the base of SMC 1 after the middle of the zone. The Corylus curve is still rising here, reaching its 60-70% TP maximum after 653 cm. SMC 2 opens at the maximum values which begin to decline after the middle of the zone. Thus SMD 2 after 653 cm would appear to chronologically correlate with SMC 2. As in SMC 1/2, there is a possibility of a small hiatus between the mineral soil and peat profiles, during which Corylus rises, and representing the Zone IV/V boundary.

The dwarf shrubs, Sphagnum and most of the herb profiles will be discussed in detail in the next chapter, as they show inter-relationships of a very local nature. However, given that SMC 2 is equivalent to the SMD 2 after 653 cm, the profiles are very similar, the very low percentages of Gramineae in SMD 2 being the only break in the pattern. Of the herbs, SMD 2 shows a little less diversity and abundance than SMC 2. The major features are similar, in particular the continuous Filipendula curve. The higher incidence of Potentilla type and Rumex acetosa/acetosella in SMD 2 can be interpreted as the relicts of the very high values seen of these species in the previous zone, confirming the local nature of herb representation in pollen diagrams. The one grain of Poterium sanguisorba in SMD 2 is of interest. It is usually confined to the late glacial or Zone IV, and to the post-Elm Decline period, and is today a plant of calcareous grasslands and limestone pavements (Pigott and Pigott 1963). It indicates a high base-status at the site and continuing open conditions. The three grains of Hedera helix that appear in SMD 2 are early indicators of the woodland. In SMC they do not occur until the next zone.

SMC 2 and SMD 2 after 653 cm have a close match. The apparently longer, more gradual invasion period for a number of species (Quercus being a good example) is due merely to the faster peat growth at SMD.

SMD 3 598.5-576 cm

Note: At 596 cm, the two series of cores for SMD were matched. Half centimetre sampling was undertaken around this point to enable as close a match as possible. Slightly enhanced Quercus, and possibly Pinus, values, and a small dip in the Calluna profile may indicate slight mismatching, erring on the earlier side. Thus, the bimodal peak of Potentilla type may in reality represent one long peak as in SMC 3. On the whole, however, the matching is smooth and presents no problems.

Correlations with SMC 3

Pine values show the same long maximum as SMC 3, but the pattern is more erratic, and the initial rise and final decline are abrupt compared to the smooth gradual profile of pine in SMC. These features are indicative of a more local profile than SMC (Janssen 1967). If, as discussed in SMC 3, pine was growing in relatively pure stands, occupying well-drained rises, then the differences between the two pine profiles can be interpreted as pine growing on the north-westerly side of the bog. Byron Edge to the north-west of SMD is a possibility, but its very proximity to the bog may indicate that it should have produced a local pattern of pine at SMC as well as at SMD. Thus, pine may be growing at a greater distance. However, the examples of Kildale Hall (Jones 1977) and Chat Moss (Birks 1964) discussed previously, indicate that such a very local pollen profile for pine is not an impossibility. The difficulty of establishing local wind patterns makes a more conclusive statement on the position of pine impossible. Birch shows similar values to SMC. Oak values are a little higher at SMD in this period, averaging 12% TP to 9% TP at SMC 3, implying the existence of oak further up the valleys on the north-west side of Soyland. However, there is no difference in the AP percentages as a whole during this zone, both SMC and SMD showing 40% by the end. The empirical limit of alder appears in this zone, as in SMC 3. Corylus too shows a decline similar to SMC 3.

The herb profile shows the greatest dissimilarity, which, of course, is to be expected. Ivy first appears in SMC during this zone and is almost continuously present from the start. In SMD, ivy appeared sporadically in the previous zone, but does not achieve such constancy as SMC. On the other hand, Lysimachia vulgaris type is constantly present in SMD 3, but only sporadic in SMC, and has obviously found favourable moist conditions around SMD peat deposits. Filipendula and Compositae are usually well represented in pollen diagrams (Andersen 1967, Godwin 1975a, respectively). Their similar, abundant profiles in SMC and SMD confirm that they show more regional than local presence. However, the herb profiles do show some matching: Trifolium, Geum, cf Ranunculus trichophyllus and Chenopodiaceae occur in this zone in both SMC and SMD. The moderately open nature of the landscape is underlined by the presence of Ophioglossum in SMD 3 and 4.

The major feature of the herbs in this zone is the Potentilla type peak, with its associated herbs, matching the peak in SMC 3 well enough to suggest that this is not a local phenomenon. However, the SMC 3 peak of Potentilla with the preceding peak of Melampyrum and Succisa is a clearer phenomenon than in SMD 3. The bimodal peak in SMD may well be a product of the marrying of the two core series. This too may have interrupted the Succisa curve which is disjointed in SMD 3.

SMC 3 and SMD 3 show a similar vegetation pattern except in the major tree species, pine, which shows a more local curve at SMD.

SMD 4 576-500 cm

Correlations with SMC 4

The correlation between the two diagrams is less apparent in this zone, SMD 4 not readily dividing into two sub-zones. The major change at the SMC 4 a/b boundary was the end of the pine maximum and its smooth decline. The more erratic pine curve of SMD shows high values until 514 cm, after which a very sudden and steep decline occurs (15-18% TP). This decline would be even more emphatic if peat growth at SMD was at the same rate as SMC: the faster rate at SMD slightly masks the comparison. There is a slight decrease before the major decline, shadowing the smooth and gradual decline of SMC 4b, but it is not a clear-cut event. As discussed, this indicates the more local nature of the pine curve at SMD, the SMC curve reflecting a wider area. Betula on the other hand has very similar values in both sites, even to reflecting a slight short peak at the beginning of the zone, after which birch declines to 5% TP in both sites. Alnus too shows a similar pattern to SMC 4, the more gradual rise of SMD 4 being attributable to the faster growing peat. Quercus stays at the 12% TO it gained in SMD 3 and does not exhibit the gradual rise in SMC 4b, reflecting the very local extension of the pine maximum. Ulmus values show the similar static pattern of previous zones in both sites until the end of the zone where it suffers a temporary decline for a very few centimetres. The significance of this, its relation to the 'real' Elm Decline, and its relation to the clearance level at SMD 500 cm, will be discussed later. This temporary loss is not seen in SMC.

Corylus profiles are similar, although the recovery which marked the SMC 4a/b boundary is not as marked. In SMD 4 hazel rises gradually through the zone, while in SMC 4, there is a small dip in values after which a definite recovery takes place. At SMD, Salix values become discontinuous at this point, and in later zones makes only a sporadic appearance, unlike SMC which shows a continuous presence throughout. This suggests that willow was growing nearer to the south-east side of Soyland.

It is during this zone that Sambucus appears sporadically at SMD. It is confined to this zone and does not appear in SMC at all, suggesting a temporally and spatially circumscribed presence. Godwin

(1975a) gives the opinion that the species involved in the postglacial is Sambucus nigra, the elder. Most records of elder are in archaeological situations, and it is always strongly associated with man, nitrogen-rich soils and rarely occurs in closed woodlands. As such, it may mark human habitation close to this side of the mire.

The profiles for the ericaceous species, grasses and sedges are markedly different in SMD 4 and SMC 4 and have a purely local significance. The herbs themselves display some notably different characteristics in this zone. The second Potentilla type peak which is extremely clear in SMC 4b has only a vestigial occurrence in SMD 4, where two levels show a small increase in Potentilla (less than2% TP). Otherwise, Potentilla occurs sporadically, unlike the constant presence in SMC 4. The associated peaks of Melampyrum and Succisa pollen are not observable in SMD 4, and Chenopodiaceae occurs only sporadically.

The woodland is represented by Hedera helix which is common for the first time in SMD. This is unlike SMC where it has been common but now declines to sporadic presence only. Humulus does not occur in SMD, but Anemone appears instead. Osmunda regalis is relatively common here, but is not seen at SMC at all. Its most likely habitat is in damp birch woods (Godwin 1975a). Polypodium vulgare shows the same increasing curve at both sites. While Pteridium follows roughly the same pattern, being high as in the previous zone, and then declining towards the end of this zone, the curve for SMD 4 is generally higher and more erratic.

The clear evidence for increasing wetness in SMC 4b is reflected in SMD in a more diffuse manner. Sphagnum values are only occasionally high, while Typha/Sparganium is absent, and Potamogeton occurs only once. In place of these, Isoetes occurs in SMD 4, suggesting the presence of pools of nutrient poor water. Cf Ranunculus trichophyllus is common during the whole of SMD 4, peaking in the middle when Isoetes is present. Caltha is present only in this zone, reflecting its occurrence at SMC 4b. Lysimachia vulgaris type is common, representing the streamside component, as do the values for Filipendula, although this latter is in fact more common in the earlier part of the zone. While it is difficult to draw conclusions from Thalictrum (which is present early in SMD 4), as its species grow in very different situations (Danielsen 1969), it is most likely to be Thalictrum flavum which grows along streamsides in wet areas. Thus, while the evidence for wetter conditions is there is SMD 4, it is more diffuse and appears to start a little earlier than in SMC 4 where all the characteristics for wetter conditions are confined to SMC 4b.

Two herb species are present which do not occur in SMC in any zone. Firstly, Rubus chamaemorus, the cloudberry, which is representative of raised or blanket bog (Godwin 1975a). Secondly, the two occurrences of cf Jasione montana are unusual here, as this species belongs to that group which normally occur as a late glacial or relict late glacial species, and then as a post-Elm Decline clearance indicator. Again, the appearance of the species in the intervening period does not necessarily point to human interference, but may merely indicate that this relatively lightly wooded area is a refugium for such species. Associations of such species need to be studied before any conclusions can be made.

SMC 4 and SMD 4 do show great similarity, as is to be expected, but SMC 4 shows a clear, decisive picture with definite episodes of vegetation change. SMD 4 is a more diffuse and blurred picture, with only vestigial indications of some of the SMC 4 episodes, particularly the second Potentilla peak. Without the comparison with SMC these episodes would not have been picked up in SMD. This may imply that changes in the vegetation pattern were occurring mainly on the southern part of the plateau, and are being reflected in the less local pollen component at SMD.

SMD 4/5 Boundary 500 cm

3870 ± 95 bc (Q 2394)

The zone boundary for SMD 4/5 was placed at 500 cm, where an obvious short clearance phase takes place. The evidence of cereal grains and high percentages of Plantago lanceolata leave no doubt of the anthropogenic nature of this clearance. However, the appearance of this clearance on the boundary creates problems in interpreting the changes in the vegetation of SMD 5, as to whether they are a result of man's activities or part of the natural vegetation succession. The boundary is dated to 3870 ± 95 bc. The corresponding SMC 4/5 boundary is dated to 4160 ± 40 bc. Taking into account the standard deviations and the subjective element in placing a zone boundary at one particular level rather than the one above or below it, these two dates are close enough to show that they are dating equivalent events, whatever the cause.

No clearance level is visible at SMC 4/5. If the clearance is confined to the pollen source area of SMD, then only the less local effects would be reflected at SMC. This, it could be argued, could have affected the positioning of the SMC 4/5 boundary, being interpreted as natural vegetation succession. However, it is difficult to conclude that the pine decline and the increase in oak and lime are the result of one small temporary clearance seen fleetingly at only one site. The pine decline is a phenomenon over most of the country and in fact the clearance at SMD 500 cm does not affect pine except in strictly temporary terms. Other direct effects are also seen as short-lived. It is therefore more reasonable to suppose that the clearance phase was part of a strategy of Mesolithic man to overcome the changing vegetation patterns, rather than as the cause.

SMD 5 500-365 cm

3870 ± 95 bc (Q 2394) - 1450 ± 100 bc (Q 2393)

Correlations with SMC 5

The pine decline at SMD is in two stages. Towards the end of SMD 4 it declines from 15% to 8%-10% TP. The lower values are maintained over the SMD 4/5 boundary before finally declining to 3-5% TP in the early part of SMD 5. At 500 cm pine temporarily declines to 4% TP, due to the effects of the clearance on all the tree values. The decline of pine in the pollen source area of SMD is not fully complete until c. 3500 bc, unlike SMC, dated to 4160 ± 40 bc, after which no local presence (ie over 5% TP) is recorded. The slightly higher values at SMD 5 may reflect vestigial presence of pine.

Oak, birch and alder show distinctly similar profiles though zone 5 at SMD and SMD, only the clearance at 500 cm, at the very beginning of the zone, disturbing the similarity. Here all SMD tree profiles decline. Interestingly, the tree profiles in this zone at both sites appear to be related to a clearance phase in SMC, at 500 cm. (Although the SMC and SMD clearances are at the same absolute level and are similar in terms of their short duration, their likeness is spurious and the two clearances are unconnected). No clearance is discernible in SMD 5, but the reaction of the trees can be seen on a larger scale than can the clearance itself. This can be observed at around 420 cm in SMD, when oak dips temporarily, although not in as clear a fashion as SMC. Alder does not show the 1 cm decline of SMC, but the general pattern of decline and then recovery towards the end of the zone is visible at SMD.

The elm curve remains at 5% TP until the end of the zone, after recovering from its short decline at the end of SMD 4. Lime and ash curves show parallel profiles with those of SMC, although ash occasionally reaches higher values in SMD (3% TP as opposed to 2% TP). The recovery of hazel at the beginning of SMD 5 after the clearance on the boundary is sharp and brief, showing hazel taking advantage of an abandoned clearing quickly before the trees re-invade. This phase is more gradual in SMC, indicating less local pollen response to events located nearer SMD. The middle and late hazel peak of SMD 5 reflects that of SMC 5.

There is an increased variety and abundance of herb types in SMD 5. Compositae types and _Rumex acetosa/acetosella_ are common. Other open and disturbed ground indicators are also present: _Trifolium_, _Ranunculus acris_ type, _Melampyrum_ and Chenopodiaceae occur. _Plantago lanceolata_ occurs sporadically through the zone. These signs of an increasing amount of open ground reflect the position at SMC 5, both recording _Plantago_ spp. Both SMC 5 and SMD 5 show an erratic _Pteridium_ curve. The two records of cf _Genista_ occur in SMD 5. Usually, this genus is incorporated within _Ulex_ type. All the species in this type are open heath or rough pasture shrubs, and these records confirm the presence of open heath, if not actually pasture areas maintained by man. The only occurrence of cf _Ononis_ here in these diagrams is unusual in pollen diagrams, the pollen being easily degradable and difficult to distinguish (P D Moore, pers comm). It occurs at only one level, 410, cm at SMD, at 2% TP, and indicates rough grassy areas (Clapham, Tutin and Warburg 1962). Its sudden appearance here in some quantity may well be associated with man's activities.

Hedera is common in both SMC and SMD 5, and is the only species, apart from the constant _Polypodium vulgare_, which represents the woodland at SMD 5. Wetland plants are relatively common in both diagrams here, and SMD 5 shows a long peak of _Potamogeton_ towards the end of the zone.

Despite the lack of clear evidence for man's activities in SMC, as there is in SMD, the pollen profiles during Soyland Moor Zone 5 show great similarity, with small deviations to the main trends visible in both.

Soyland Moor Site A

This core was taken at the southern edge of the bog. The construction of the A58 between the edge of Soyland Moor and Blackstone Edge destroyed the extreme edges of the peat deposits, but the SMA cores were taken as near as possible to the edge of the undisturbed deposits. The peat depth was 310 cm, with the mineral soil samples reaching to316 cm.

SMA 1 316-311 cm Pre-peat material

The pre-peat material indicates a wooded environment. The high amounts of Gramineae, and ericales other than *Calluna* may be contamination from much older deposits, these high profiles having no correlation with the peat deposits of SMA or the other peat cores.

Description

Alnus and *Quercus* values maintain 10-15% TP, while *Betula, Ulmus* and *Pinus* values show an average 4% TP. *Tilia* pollen is present in small quantities (less than 1% TP). *Corylus* dominates the spectrum at 30% TP. The values of Gramineae and ericaceous species are high at 15% TP each.

Potentilla type is common, with *Hedera helix, Filipendula* and Umbelliferae occurring often. Most of the herbs occurring once only are disturbance or open indicators. Both the *Pteridium* and *Polypodium vulgare* curves show constant low percentages.

Interpretation and Correlations

SMA 1 shows a correlation with the SMC 4/5 boundary. Alder and oak are present with high values but not as high as the full extension of both species in SMA 2. *Pinus,* while not showing the high values of the pine maximum, has not yet fully declined. *Tilia* is present in minute, but consistent amounts. *Fraxinus* has not yet appeared.

Hedera helix is the only species representing the woodland, and open species are more common, such as *Valeriana dioica* (Mamakowa 1968) and *Galium* type.

Hazel and heather are the dominant species in this zone, echoing most of the other diagrams in the mosaic of open woodland - heath - grassland - hazel scrub. When compared to SMC, this zone can be dated to 4400 bc. At SMC and SMD the long *Pinus* maximum delays the full expansion of deciduous forest, and it is to be assumed that this was also the case at SMA.

SMA 2 311-377 cm

Correlations with SMC and other diagrams

The *Betula, Ulmus* and *Alnus* curves in SMA 2 are identical with the early part of SMC 5, before the clearance phase at 500 cm. However, the lower *Quercus* curve compares with the later part of SMC 5. The major difference between SMA 2 and SMD 5, however, is the aberrant *Tilia* profile. In the early part of this zone, *Tilia* has values of over 5% TP, unlike the other 3 diagrams, where values of 3% TP are the norm. AT 292 cm, the curve drops suddenly to 3% TP. The only change associated with this decrease is a slight increase in

Alnus. It is unlikely that an increase in alder locally would depress the Tilia curve selectively, so that it can be accepted that there was a real decline in lime, and that it was a local event. The high values at SMA must indicate that lime was growing very near to the site, possibly beyond the south-east edge of the mire at the top of the River Ryburn valley, signifying that Tilia reached higher altitudes than was previously thought. Fraxinus appears at the base of this zone and rises to 2% TP, as in SMC 5. Acer also occurs in the early part of SMA 2, reflecting the single occurrence of SMC, which was a little earlier, at the end of SMC 4b.

Corylus shows a reasonably similar profile. Ilex and Sambucus occur in SMA 2, showing a correlation with SMD rather than SMC. The low pollen production of these two species could account for their absence in SMC. The high incidence of Sambucus at SMD indicates that it was growing locally on the north west side of the bog. The single occurrence at SMA could either reflect that local presence, or indicate that elder had spread to the southern edges of the bog.

The ericales show a similar pattern to the early part of SMC 5 (the peak after the SMC 500 cm clearance is not present). Calluna as usual shows highly local patterns. Conversely, Gramineae and Cyperaceae show curves very similar to those of the early and central part of SMC 5.

The wet conditions seen at SMC 5 are echoed at SMA 2, confirming, with SMD, that this is not a very local event. SMA 2 has high Sphagnum values and the sporadic occurrence of Typha/Sparganium. The wetland herbs are not diverse at SMA 2, but the peak of Potamogeton correlates with that of SMD 5. This is lacking at SMC 5, indicating pondweed's local presence in pools on the bog or in nearby streams.

The commonly abundant herbs of Potentilla type and Filipendula are here in SMA 2. The woodland is represented by abundant Hedera and the occurrence of Humulus. This is similar to SMC, although there, Humulus occurs a little earlier. Osmunda regalis is recorded. It is missing in SMC but present in SMD. The open and disturbed ground indicators are generally common to SMC and SMA. Compositae and Rumex acetosa/acetosella are abundant, and Chenopodiaceae, Urtica and Rumex obtusifolius type occur in both. Plantago lanceolata has a sporadic appearance in both diagrams, indicating increasing areas of open ground. Malva type, signifying disturbed ground, occurs at SMA 2, its only occurrence in the four diagrams. Three unusual herb types are recorded in SMA 2 (see Appendix III for identification). Firstly Mercurialis cf annua occurs, in a similar position to its occurrence in SMC 5. Cf Geranium and cf Erodium occur in SMA 2, their only recorded occurrence in the four Soyland diagrams. All three may be indicative of anthropogenic disturbances.

Apart from the unusually high Tilia values and its sudden, early decline before the Elm Decline, SMA 2 correlates well with the lower part of SMC 5, up to the clearance dated to 2915 bc. This correlation may indicate that the Elm Decline at SMC is at this earlier date rather than the 2615 bc postulated. This will be discussed in detail later.

Soyland Moor Site B

This core was taken 70m from site A in a north-westerly direction, towards the deepest part of the bog. The base of the peat occurs at 430 cm, with the mineral soil sampled to a depth of 455 cm. The softness of the pre-peat material allowed their deeper sampling, but also made imperfect cores. Thus sampling was at greater intervals.

SMB 1 455-432 cm Pre-peat material

This material showed the greatest mixing and contamination of all four cores. Quercus and Alnus have a contemporary rational rise, and the curves for Betula, Pinus and Corylus are highly variable. There are also a number of features which could relate to the late glacial or early postglacial, in particular the occurrence of Ephedra which does not occur after Zone III in Britain (Godwin 1975a). The 5% TP values of Rumex acetosa/acetosella, the high values of Gramineae (up to 50% TP) and the small curve of Empetrum all suggest a much earlier date than do the trees. Thus, this material is not useful for interpreting the pre-peat vegetation.

SMB 2 430-409 cm

Correlations with SMC, and other diagrams

Despite the greater depth of SMB (430 cm) compared to SMA (310 cm), and its position nearer the centre of the early peat accumulation, SMB appears to have started a little later than SMA, and covers only the end of the Atlantic period. The Tilia and Fraxinus curves are well established at the peat base, which indicates a later initiation date than SMA. The occurrence of these two species in the pre-peat zone of SMB 1 may confirm that peat started forming only after lime and ash were established in the region, although the highly mixed nature of SMB 1 makes this evidence unreliable.

Given the very short profiles it is difficult to correlate SMB with the other diagrams with any degree of accuracy. The low values, comparatively speaking, of Quercus and Alnus may suggest a correlation with the later part of SMC 5, ie after the clearance at 500 cm, but the Corylus curve is more similar to the early section.

The sole identifiable feature of this zone is the continuous abundant presence of Potamogeton, now seen in each diagram at this point, except SMC. This may not be a resonable correlation, Potamogeton being a very local species which would commonly occur in pools and streams. If the correlation can be made, then, by extrapolation from SMA 2, this zone most likely relates to the earlier part of SMC 5.

Conclusions

The period covered by these four diagrams is from the early postglacial to the Elm Decline. This early Flandrian period saw rapid climatic change from the sudden warming up at 8300 bc through the dry Boreal and into the warm and moist Atlantic period (Lamb 1977). Vegetation changes were concomitantly rapid. Further up the chain, animals, including man, had to adapt to the environmental changes in order to survive. The rising of the sea level continued through most of this period, the breaching of the Channel occurring at the beginning of the Atlantic (Godwin et al 1958). While this affected vegetation in terms of migration from the Continent and in increased oceanity, the major effects were felt by man, who now had to live in a more circumscribed environment. As discussed in Chapter 2, the uplands of Britain were the more favoured areas in the initial post-glacial period, and the area around Soyland was particularly intensively used.

Soyland Moor occupies a central position in Britain and its pollen data reflects this, showing a mixture of sometimes apparently contradictory characteristics. Its height of 380 m is today classed as highland (eg Barnes 1982) and certainly altitude has a large role to play in determining the vegetation patterns and sequence. However, it is by no means as high as areas of the northern Pennines and the Lake District (I exclude Scotland here because of its very different postglacial vegetation sequence), and often 'southerly' or 'lowland' features are to be found at Soyland. The high *Tilia* levels are unusual at this altitude and this far north, and *Ulmus* values too are relatively high. Yet the arboreal pollen percentages are never high, remaining at around 40-50% TP during the Atlantic period, indicating woodland is never closed. Clearance by Mesolithic groups may well have had a great effect on the maximum extent of the woodland, and thus explain this apparent discrepancy.

On the whole, Soyland Moor displays 'westerly' affinities, having more in common with the Lake District than the North Yorks Moors, for example. In particular, the pine profiles indicate this. Pine values are high and rise relatively late compared to the south and east. Godwin (1975b) suggests that there are two separate populations of pine in Britain, one in the south and east, the other in the north and west. If this is so, the Soyland population belongs to the north and west. The long pine maximum and the slow alder rise are more unusual features and suggest comparisons with Irish and Scottish sites. However, Godwin (1975a) explains the Irish phenomena by widespread mire formation and the Scottish ones by the inability of alder to successfully invade so far north. Around Soyland, mire formation was no more widespread than in other upland areas of Britain, and alder had no difficulty in initially establishing itself, so such affinities are spurious.

The high *Corylus* values, both initially and throughout the profile, could also indicate 'westerly' affinities (Smith 1972), but this is a common phenomenon in all upland areas of Britain, and merely indicates that, once established, hazel can hold its own at altitudes where trees cannot establish closed woodland, or where man is maintaining open areas.

Robinson's Moss, 20 km to the south (Tallis unpublished), shows much the same sequence as Soyland, perhaps indicating that the vegetation changes shown are regionally applicable. The long pine maximum is present, with alder rising while pine is still dominant. This unusual feature may well be the hall mark of the central Pennine area, and runs counter to the hypothesis that pine does better on the limestone (Turner and Hodgson 1979, Bartley et al 1976), although a monothetic explanation is probably not applicable to pine, a species that does well on a variety of substrata. The associated delayed full expansion of oak is also seen at Robinson's Moss. Birch shows similar equivocal low values and the profiles show very similar curves at both sites. The initial decline is in neither case associated with the rise of pine as it is in many other sites. Indeed, during the pine maximum, both birch profiles show a temporary rise in values before declining to a new low which is maintained until the Elm Decline. The unusually early occurrence of high _Calluna_ values, before pine values rise, is also a common feature (Zone VIIa or post Elm Decline is a more usual date for the widespread occurrence of _Calluna_).

These common characteristics indicate, at the very least, that the Soyland diagrams show more than local variations, and, at most, that these features have a regional significance for the whole of the central Pennine millstone grit plateau. Two sites are not enough to allow a categoric statement and merely pave the way for confirmatory evidence. This may take time in coming forward, as peat development at such an early stage in the postglacial is unusual in the central Pennines.

CHAPTER 5

THE SPATIAL ANALYSIS OF THE POLLEN DATA WITH REFERENCE TO THE MIRE VEGETATION AND THE SURROUNDING WOODLAND

The interpretations of the last chapter are to a large extent based on comparisons with other sites. The intra-site comparison was more concerned with similarities in the pollen data than differences. In this chapter the differences between the four diagrams will be analysed, in spatial rather than chronological terms. The surface vegetation of the mire will be discussed, as will the mire's development in time and space.

Soyland Moor is a good site for analysing local vegetation changes. Its small size for much of its history allows for greater representation of local pollen (Janssen 1966) and the analysis of four cores from one site allows for spatial and directional conclusions (cf Turner 1975). Episodes which are sharp and clear in one diagram may be blurred or occur only in vestigial form in another. Other episodes or trends are clear in all diagrams, but may show different percentage values. A further advantage is the greater amount of information on local vegetation changes. Very small changes in a single diagram often have to be ignored partly because not enough information is available to interpret them, and partly because they may merely represent pollen count fluctuations. On studying more than one diagram, small fluctuations could occasionally be seen to be echoes of a clearer episode from another diagram. This is of particular use when looking for evidence of man's activities but also for the elucidation of the developmental sequence of the mire itself.

A similar technique on a much larger scale was undertaken by Turner and Hodgson (1979, 1983), where large numbers of sites across the northern Pennines and County Durham lowlands were statistically analysed for variations in pollen values. One of the major contributions of these studies was the conclusion that spatial variation was greater than temporal variation. What the smaller scale study at Soyland shows is that local vegetation is just as variable as that of the region. The statistical tests applied by Turner and Hodgson would hopefully be applicable here, but could be of limited value. To be accurate, a greater number of radiocarbon dates would be needed to ensure comparison between similarly dated levels. Also, much of the local variation is based on small changes in values, and also on the species which produce little pollen, which would not register as statistically significant. Weighting the relevant figures is possible, but pre-supposes their significance, and thus the tests merely confirm the subjective hypothesis.

The interpretation of the data in spatial terms is heavily dependent on modern pollen research. Tauber (1965) produced the classic basis for interpretation, which is worth repeating here as his model is extensively used. In lakes and mires, the woodland pollen content is made up of three components: pollen carried between the trunk spaces of the woodland (locally derived); pollen carried by air currents above the tree canopy before falling (less local and regionally derived); pollen carried higher above the canopy which stays in the atmosphere until rain brings it down (regionally and longer distances). The proportion of each component within a deposit is dependent on two factors; the size of the catchment (Tauber 1965), and the distance of the woodland from the catchment (Birks and Birks 1980). The smaller the catchment and the closer the woodland, the higher the proportion of local pollen. This hypothesis was extended by Janssen (1966) to include pollen produced by the bog itself. Again, the smaller the bog, the higher the proportion of pollen from the bog vegetation itself. The nomenclature for these four components changes from author to author. Throughout I have used 'very local' for the bog vegetation, and that on the edges of the bog (Janssen's 'local' pollen), 'local' for Tauber's 'trunk space' and Janssen's 'extralocal' component, and 'regional' for Tauber's 'canopy' and 'rain-out' components.

There are of course many factors which may interfere with this model, such as the density of the vegetation, climatic conditions and local topography (Cundill 1979, Wright 1952 and 1953); treeless landscapes may need different and much more sophisticated models (Pennington 1980). However, experiment has confirmed the validity of the basic tenets (eg Tauber 1967, Janssen 1967, 1973, Caseldine and Gordon 1978), and generalised figures have been estimated. Thus a small mire 100-200 m across will receive up to 80% of its pollen from very local and local sources (Tauber 1967b). These figures are of particular relevance to the Soyland diagrams as, for most of the pre-Elm Decline period, the mire was under 200 m in diameter. A further discovery was the very rapid exponential fall-off of pollen from its source, within 100-200 m depending on the strength of the source (Janssen 1973).

Turner (1975) used this model extensively in her 'three-dimensional' diagrams from lowland Ayrshire, as did Tallis in the central Pennines (Jacobi et al 1976). Interpretation was based on the distance of the contemporary cores from the edges of each mire. For most of the relevant time period at Soyland, the mire bog was very small, and was represented at SMC and SMD only. Both core series are, therefore, near the edge of the mire and reflect mostly very local and local pollen input. Only with the spread of mire across SMA and SMB, can local and regional pollen input be distinguished. Thus only the very end of the sequence can be treated in the same way as the Ayrshire sites.

Some of the other conclusions of modern pollen research are of general application. Janssen (1973) has pointed out that very local pollen can often be identified by their very erratic curves. This is due in part to the greater fluctuations of local pollen production (Mandrioli et al 1980, Litchi-Federovich and Ritchie 1965) which are smoothed out only with distance. Also, the fluctuations represent a closer reflection of small changes in vegetation patterns. Dimbleby (1957) suggests two rough rules of thumb to further identify local pollen: values greater than 2% TP indicate presence within 300 m of the site, and any values higher than 5% TP are indicative of the immediate surroundings of the catchment (these figures are referable to the large pollen producers only). Both Tinsley and Smith (1974) and Litchi-Federovich and Ritchie (1965) state that arboreal pollen values over 50% TP represent open woodland within 100 m. Woodland herbs and open ground herbs in open woodland habitats often do not travel far enough to be deposited in mires (Janssen 1967), and pollen from some low-growing herbs do not travel further than a metre from the parent plant (Moore 1976a). The exceptions here are open ground herbs in a treeless area, where dispersal is greater, and some herbs, especially the taller ones such as *Filipendula* and *Urtica,* do disperse well even in forested environments (Andersen 1970).

There are two major problems unresolved. Firstly when both population size and distance from the catchment is unknown, interpretation may be difficult: do the values represent a smaller population near the catchment or a larger one at a greater distance (Jacobson and Bradshaw 1981)? This may be partly resolved by a consideration of ecological requirements and associations, and by such 'three-dimensional' diagrams as presented here. Secondly, does a change in values represent a change in the population size or reflect changes in dispersal only (Tauber 1967a)?

Despite these unresolved problems, modern pollen representation studies hold the best hope for a correct interpretation of diagrams in spatial terms, and this type of information has been used in discussing the Soyland Moor diagrams. The work undertaken in Nidderdale was particularly useful here (Tinsley 1972, Tinsley and Smith 1974), as the environment is similar to Soyland. Both comprise a millstone grit plateau, with the Nidd-Laver interfluve being on average slightly lower, with a maximum height of 366 m. Both plateaux are cut by deeply incised valleys, and have similar present day vegetation.

Local Mire Vegetation and Development

Here, I shall look at the very local pollen profiles, in particular at *Calluna,* Gramineae, Cyperaceae and *Sphagnum. Sphagnum* curves have to be used with caution, as its sporing capacity changes dramatically under different circumstances. There is some indication (Conway 1954, Tallis 1964c) that large increases in *Sphagnum* spores are correlated with wetter phases in mires, shown by bands of *Sphagnum* macro-remains in the stratigraphy. However, Tinsley (1972) shows that modern sporing does not correlate with any consistent change in the environment. The difficulty of consistently identifying *Sphagnum* spores to species level compounds the problem.

Firstly, the vegetation prior to peat initiation is discussed. The vegetation of the topogenous mire, that is SMC and SMD until c. 3000 bc, forms a separate section, and, lastly, the spread of the peat as blanket bog, in the immediate pre-Elm Decline period, is looked at to determine the effect that this had on the mire and surrounding vegetation.

The pre-peat vegetation

The hypothesis that the Soyland Moor peat deposits started life as a topogenous mire, filling in the shallow valley at the confluence of two streams which possibly formed a small lake, is corroborated by the pollen data. Both SMC 1 and SMD 1 exhibit many streamside elements. In SMD 1 the pre-peat vegetation is indicative of an early period, either late Zone III or early Zone IV, and its late glacial elements have been discussed in the previous chapter. SMC 1 represents Zones IV and V, with only cf *Oxyria digyna* and *Lycopodium alpinum* as definite late glacial relicts. It is possible to interpret these late glacial and early postglacial elements into the community they represented on site. Salix carr seems to have developed at the confluence of the two streams (P D Moore, R T Smith, pers. comm.). The high Salix and Filipendula values and the continuous record of Ranunculus acris type all indicate this development, as does the occurrence of Chenopodiaceae. Caseldine and Gordon (1978), discussing modern pollen rain from willow carr, find very similar pollen types and values. Given that Salix was growing on the site itself, it follows that most of the pollen input would have been very local, the shrub cover masking out much local and regional pollen (Janssen 1967, Andersen 1967). Therefore the large birch values may well represent trees growing on site, although some may still represent long distance transport (Moore and Bellamy 1974). On damp soils, Betula pubescens is the usual birch species (Godwin 1975a). The high values that persist into SMD 2 suggest that in the very early stages of mire development, birch remained either on the peat or around the edges. By the time peat had started forming at SMC, birch had retreated from the peat edges onto the plateau top.

The willow carr was surrounded by marshy ground, probably between the streams and the carr itself. The high Polypodiaceae values could represent the fern element within the marshy tall-herb community (or a dwarf shrub-fern vegetation as postulated by Tallis (1964a) for Snake Pass). In the initial stages, this community has some rich fen affinities. The high Cyperaceae of SMD 1 is consistent, and may represent Cladium (Holdgate 1955). The high Potentilla is interpreted as possibly representing the very local dominance of Potentilla palustris, the marsh cinquefoil, which is common in the less productive fens (Holdgate 1955). P. palustris is in these situations often associated with Galium, which also occurs in SMD 1. The presence of Scabiosa columbaria and Polemonium caeruleum in SMD 1 confirm the relatively high base status of the soils. Today, these species only occur on limestone (Clapham, Tutin and Warburg 1962). The later period of this fen carr, as represented by SMC 1, shows that these high base-status associations had diminished, indicating early leaching in the wetter parts of the mineral soil. Millstone grit soils are not base rich as a rule, the early bases having derived from the breaking up of bedrock and the formation of raw soils by periglacial action (Pearsall 1968). The high Cyperaceae values of SMD 1 have declined, and Potentilla type occurs only sporadically. The basal sample of SMC 1 only shows the last vestiges of any rich fen affinities. Also the very low values of ericaceous species and Empetrum in SMD 1, which indicate non-local presence, increase enormously in SMC 1, especially the ericales other than Calluna. These species represent the growth of acid heath, and could be associated either with the large areas covered with Sphagnum, where Erica tetralix and Empetrum nigrum are common elements, or as dryer heathland under the open birch woodland, where either Vaccinium spp. or Erica tetralix could be dominant (Bell and Tallis 1973).

The extremely high Sphagnum throughout SMD 1 and at the base of SMC 1 is interesting for as yet there is no peat growth. However, willow carr can be associated with base-tolerant Sphagnum spp. such as S. fimbriatum, S. teres and S. squarrosum, all of which today can occur in carr or wet woodland (Dickson 1973).

Thus, the pre-peat period at Soyland Moor supported Salix carr with its initial rich fen affinities, which gradually declined over a long period of time as leaching took place, and heathland became more important.

The inter-relationships of the mire vegetation at SMC and SMD

Until the immediate pre-Elm Decline period, the peat deposits at Soyland are topogenous, and are represented only at SMC and SMD. The most noticeable feature of the local pollen profiles of the two sites is their considerable variation, and occasional total dissimilarity, indicating that they are representative only of very small areas of mire vegetation. Too often, the sequence of the surface vegetation of an entire mire is extrapolated from one diagram. A core taken from the middle of even-aged blanket or large raised bog will be more representative of the surface vegetation as a whole than here, but it will still be enhanced by very local dominance patterns.

I shall first discuss the very local pollen types separately, comparing SMC and SMD before discussing their interactions on the mire surface. The major complicating factor is that, apart from Sphagnum, which has its own interpretation problems, all these species may form part of the dry land pollen as well as that from the mire itself.

The mire vegetation species

Ericales: In both SMC and SMD (apart from SMD 1) the ericales other than Calluna show initial high values, although the Calluna values are still high enough to suggest local presence. Taking all the ericaceous species together they suggest that local leaching has taken place, most likely at first down the stream banks and later down the slopes to the peat edges. However, Godwin (1975a) suggests that in cases of early leaching only low values of ericaceous pollen are present. The high values here can be explained by some ericaceous cover on the mire itself. However, this would not explain the very high values at SMC 1, which has a very local pollen input and would not reflect the mire surface which is present only at SMD at this early period. Thus the ericales at SMC must represent the dry land pollen of heath communities which may have taken early advantage of leached areas by spreading out from the mire surface. The decline and virtual disappearance of the ericales (Vaccinium and Erica) earlier at SMD than SMC suggests that these species did not invade as a ground cover under the encroaching pine stands. Pine shows more local presence at SMD and it is there that one would expect the ericales to persist. The extension at SMC suggests the local persistence of small open Vaccinium or Erica heath communities to the south-east of the bog right through to the Elm Decline, although against increasing odds.

The Calluna profile itself is more difficult to interpret. In general, SMD shows a less erratic Calluna profile than SMC with generally higher values, either indicating a local rather than very local profile or that a more stable and extensive area of Calluna remained on the northerly side of the bog. The SMC Calluna curve shows many peaks and troughs, including a long period when it was no longer present around SMC (5022 ± 40 - 4390 ± 50 bc). At SMD at this period, Calluna declines but still shows a considerable local presence. Despite the differences, the two Calluna profiles of SMC and SMD show a similar pattern of rise and fall. The differences can be explained in terms of the very local pattern of Calluna growth on the mire surface and around its margins. The general similarity, however, suggests that a less local pollen input (ie off the mire surface) is also well represented. Thus, if the dry land Calluna input around SMC decreases drastically, a smaller decrease is registered at SMD.

Gramineae: Gramineae pollen is wind transported and it would be expected to show a local or regional input. However, the very different curves for grasses at SMC and SMD indicate that the majority of grass pollen input is from the bog surface vegetation itself, despite the low values recorded, in which case the species involved is likely to be Molinia caerulea. The low values therefore must truly represent the small population of grasses on Soyland. At SMD, values are generally below 2% TP, with only one period of high grasses which peaks at 15% TP, occurring at the Potentilla type peak. At SMC, a peak occurs in a very similar position. This simultaneous occurrence, and the strong possibility of large-scale clearance at this time imply that dry land pollen is represented here. After this period, the grass pollen at SMC shows at least four further peaks, none of which are even echoed at SMD, and it is difficult to state whether it represents the bog surface or dry land communities. This is a common problem, the habitat range of grasses being so wide, and the pollen identification being confined virtually to family level. The general record of Gramineae is too low to suggest extensive presence on the bog surface, which may indicate that the infrequent SMC peaks are representative of dryland pollen. The hints of more open ground around SMC than SMD pollen source area may confirm this.

Cyperaceae: Cyperaceae suffers from the same problem as the grasses, namely, that distinguishing the pollen below family level is mostly impossible. Litchi-Federovich and Ritchie (1965) state that 1-2% TP values indicate that Cyperaceae is present from dry-land pollen rather than from the mire vegetation. The low values for most of the two diagrams and the slightly better correlation than that of the grasses between SMC and SMD may suggest that this is the case at Soyland. However, it is unlikely that it ever disappears entirely from the mire surface and the occasional occurrence of Rhynchospora alba confirms this. Cyperaceae pollen is highly degradable and can be under-represented in the peat deposits, showing low values even when it forms substantial communities on the mire (Rybnickova and Rybnicek 1971).

Sphagnum: The loss of the very high Sphagnum at the end of SMC 1 and the SMD 1/2 boundary may seem odd at a period when peat had or was just about to form on site. However, its loss can be explained in a number of ways. Firstly, this period is correlated with a general drying out under a continental type climate (Lamb 1977), drastically reducing the number of habitats for Sphagnum across the whole of the landscape (despite the onset of peat accumulation at SMD and SMC, which is due to purely local hydrological conditions). Secondly, the initial phase of the mire may not have been dominated by Sphagnum, although there are no signs of any other mire builder at this period. The high ericales values may indicate a relatively dry surface but the actual growth of the peat is fast, which should indicate it was continually waterlogged. Thirdly, the Sphagnum species present on the bog may well be different from those in the early, base-rich period, and may have different sporing characterisation. SMD 2 and late SMC 1 show very erratic Sphagnum values which make interpretation of its position difficult.

Zones 3 and early 4 have low Sphagnum values, but SMC 4b shows a sudden and consistent increase, as well as an increase in not only mire herbs, but aquatic and streamside herbs too. Together, these indicate higher water tables, which is consistent with evidence from the rest of Britain. Caltha is present for the first and only time. Typha/Sparganium becomes common, as does Lysimachia vulgaris type. SMD 4 shows these characteristics in a more subdued and erratic fashion: Sphagnum increases at occasional levels, Lysimachia is common, but less abundant than in SMD 3. The appearance of Caltha echoes the SMC 4b, but Typha/Sparganium does not even appear in SMD 4 after consistent presence in earlier zones. Taking the evidence together, it suggests that the north-west side of the mire was wetter in the earlier period, which is confirmed by its earlier peat initiation and faster growth. By zone 4, however, there are more indications of increasing wetness at SMC than SMD, perhaps indicating that peat was now growing faster at SMC. However, there are no indications of this within the peat stratigraphy.

The inter-relationships of the mire vegetation

SMC shows a clear cyclical relationship between Calluna and Sphagnum. In the wetter periods, Sphagnum dominates the bog, and sometimes Calluna disappears from the bog altogether, as in SMC 4b. In the drier periods, Calluna may then become dominant, although it is doubtful that Sphagnum ever disappears, as this bog remains relatively wet throughout. Calluna is most vigorous on dry heaths (Gimingham 1975), but can still be an important component of wetter bogs (Gimingham 1960). SMD shows a similar inverse relationship, but in a much more muted fashion, and occasionally departs from it altogether. In SMD 2, both Calluna and Sphagnum have very erratic curves, which may suggest Calluna was generally dominant at the the edges of what was a very small bog at that period, with Sphagnum in the wetter centre. The sudden and transient shifts in the dominance pattern may then reflect minor and very local hydrological changes. In later periods, Calluna and Sphagnum maintain more equable values over long periods of time, and, although Calluna decreases in late SMD 4, it does not disappear from the mire surface as at SMC 4b. Correspondingly, Sphagnum does not peak consistently. As Sphagnum is calculated outside the pollen sum, this must be a real and not a percentage phenomenon.

In SMD 3 and early SMD 4, the cyclical relationship seems to be between Sphagnum and the grasses and sedges rather than Calluna. As Sphagnum declines gradually, both the grasses and sedges increase slightly, until the increase of Sphagnum during late SMD 4. No such relationship can be seen at SMC, nor do the grass and sedge profiles match as at SMD. While this may seem to indicate that at SMD Calluna is representing heath communities off the mire, the generally low values for Sphagnum, grasses and sedges mean another species must be dominating the mire, ie Calluna. Thus the stable values for heather are more indicative of widespread Calluna on this side of the mire.

Hazel is a complicating factor to this interpretation of the mire surface vegetation. While it is not suggested that hazel was growing on the mire, it too shows an inverse relationship with *Calluna*. Therefore, if *Sphagnum* were ignored, these profiles could be interpreted as a cyclical dominance pattern around the mire between heathland and hazel scrub. This is acceptable in ecological terms, as *Calluna* values increase during clearance phases, and peak afterwards, before *Corylus* values increase. The slight peaks in the *Betula* profile between the peak in *Calluna* and the rise in *Corylus* values may provide the intermediary stage between acid *Calluna* heath with incipient podsol soils, and a mull humus soil which would characterise hazel scrub (this is discussed in more detail in Chapter 6). The relationship between *Sphagnum* and *Calluna* would then be spurious in community terms.

There is one further possibility, that a proportion of the Coryloid pollen is referable to *Myrica gale* rather than *Corylus*. If this were so, then the inverse relationship would be between *Calluna* and *Sphagnum* with *Myrica*, the latter also increasing as the mire became wetter. Most researchers agree that *Myrica* was not important until after the Elm Decline (Godwin 1975a, Turner and Hodgson 1983), but, at Soyland, there are many features which do not generally appear until this later period (*vide Calluna* itself and many of the herb pollen types). While the similarity of the Coryloid profile at Soyland to most other British sites could argue against a strong input of *Myrica* pollen, it also argues against *Corylus* having a seral role after what are presumably local clearances (especially in having an inverse relationship with *Calluna* which does not occur in quantity at most other sites of the period).

To summarise, the inverse relationship between *Calluna* and *Sphagnum*, and *Calluna* and Coryloid values, can be explained in three ways:

1. A dominance cycle on the mire surface based on wet and dry periods. The Coryloid profile is then spurious. Possible arguments against this are the lack of such a clear relationship between *Calluna* and *Sphagnum* at SMD, and the lack of evidence for other indications of wetness to accompany the *Sphagnum* spore increases (apart from during SMC 4b).

2. The successive phases of post-clearance colonisation. *Calluna* invades, followed by *Betula*, then *Corylus*. *Sphagnum* is then spurious. The similarity of the Coryloid profile to other sites might militate against this.

3. *Myrica gale* is significantly present in the Coryloid values. This has the advantage of allowing all three profiles into the explanation, with *Calluna* dominating the mire surface in the dry periods, *Sphagnum* with *Myrica* in the wet periods. Again, the similarity of the Coryloid profits to the normal sequence argues against this.

On the whole, the clearance explanation seems to be most appropriate for these diagrams at least, and possibly for many others. The *Sphagnum* profile would then be explained by its eclectic sporing characteristics.

The spread of blanket bog

In the immediate pre-Elm Decline period, the extent of the mire increased enormously. Peat had by now filled the shallow basin and had begun to spread laterally. However, the spread of peat across the plateau is not entirely connected with the growth of the topogenous mire, as peat forms at SMA prior to SMB. This occurred despite SMB's position closer to the topogenous mire and despite its greater depth of peat deposit. Thus, the formation of the blanket bog at Soyland was a part of the general accumulation of blanket bog across the area at Soyland mire. In the southern and central Pennines as a whole, the dates for the formation of blanket bog range between 5000-3000 bc. The high plateau sites show the earliest dates (Tallis 1964a). At Soyland, SMA started to form between 4000-3000 bc. The low pine levels, particularly those in the pre-peat material, date it to after 4000 bc (using the SMC dates), but the low and steadily increasing ash curve indicates a pre-3000 bc date. SMB, as mentioned above, has an even later date than SMA. Given the altitude of 389 m, these dates for the spread of blanket bog are relatively late. Thus, Soyland Moor peat formation as a whole shows unusual features, both in the very early inception of the topogenous mire, and in the late spread of blanket bog. However, this late spread is still well within the general range in both the Pennines and the uplands of Britain as a whole. The North York Moors, for example, show a similar, late pattern (Cundill 1977). The causes of the waterlogging leading to the peat formation may simply be a result of the increasingly wet climate during the period between 5000-3000 bc, but may be complicated by deforestation by the Mesolithic populations (Moore 1975a). The bog surface throughout remains wet as shown by the generally high *Sphagnum* at all four sites. *Potamogeton* shows a continuous presence in the pre-Elm Decline period in sites A and D, while SMB shows a definite peak. Its absence at SMC underlines its very local distribution around the other sites. If these three phases of *Potamogeton* can be correlated, then this period represents an increase in the incidence of pools on the bog surface. Wetland and aquatic herbs are abundant also around this period.

Calluna values increase at SMC and SMD with the extension of the mire, and are high also in SMA and SMB. For the first time, *Calluna* curves are consistent from diagram to diagram. These smooth curves of high value signify that heather was dominant over large areas of the mire, indicating that podsolisation was occurring across the plateau. This is confirmed by the decreasing arboreal and *Corylus* values during this zone. The indications of increasing wetness may well suggest one of the causes of podsolisation. It is also possible that, given the enormously increased area of the bog, there were drier areas of the mire surface dominated by *Calluna*. If that were the case, however, then higher *Calluna* values should be registered at SMA and SMB, as they are nearest to the edge of the bog and thus show the higher values of local pollen (Janssen 1968). SMA and SMB do not show these higher values, confirming that the majority of the *Calluna* values represent dryland heath. it is during this period that *Calluna* values increase enormously at most British sites. Thus, despite the early rise in values, the Soyland *Calluna* profile follows many other sites in showing widespread heathland development at this period.

Both the Gramineae and Cyperaceae values are generally more important at SMA than the other sites, implying, using Janssen's

(1966) argument, that these represent plants on the mire surface itself, or around its edges.

This period is basically regressive, with the spread of peat and the decrease in tree and shrub cover. Calluna values increase substantially on the acidic soils, and thereby help to initiate the process of podsolisation which quickly became irreversible (Handley 1954).

The Extent, Distribution and Composition of Woodland around Soyland Moor

The extent of the woodland

At SMA 1, SMD 1 and early SMD 2, the extent of the woodland on the plateau top is difficult to judge because of the very local pollen input which shows high birch and willow as well as high herbaceous pollen. Throughout upland Britain, Zone IV represents an essentially open landscape with copses of trees in favourable positions, and the over production and long distance transport of birch and pine pollen is a strong feature of diagrams from this period (Birks and Birks 1980). Thus, as previously discussed, willow and birch are growing on site while the surrounding landscape remains open, as shown by the high herbaceous values. At the end of SMC 1 and during SMD 2 the tree values decline to 10% TP, which is perhaps a truer indication of the position of the woodland in the landscape. According to Tinsley and Smith (1974), arboreal values below 15% TP indicate a relatively treeless landscape. After birch was no longer growing on the mire site itself, it probably formed small copses on the wetter areas of the plateau, by which time pine had also arrived in the immediate area. The high hazel values (60-70% TP) suggest hazel was the dominant cover over Soyland Moor. When not an understorey shrub, hazel will form a small tree, but will also produce up to four times as much pollen than it does as a shrub (Godwin 1975a). However, even if the figures are corrected, hazel was still the dominant woody component of the vegetation at this time.

By 5025 ± 40 bc, the steadily increasing amount of woodland pollen has reached its maximum, averaging 40-50% TP at both SMC and SMD, rarely reaching beyond 55% TP. The mire was still small at this point, so these figures suggest open woodland within 100 m of the site (Tinsley and Smith 1974). The date here corresponds to the Atlantic period, when maximum forest cover was reached over much of Britain, as here, but also when the woodland was dominated by oak, lime and elm, and alder and ash are also important. Here, pine was still at its maximum and the deciduous elements, while present, were still in a subordinate position. This suggests that a delaying factor, such as altitude, migration rates or unknown local conditions, produced an inertia within the established woodland which still showed a Boreal rather than Atlantic composition.

When comparing these maximum woodland values to other, similar sites, the figure of 40-50% TP is common. Many North York Moors sites have similar values (eg Fen Bogs (Atherden 1976)), as do Valley Bog in the north Pennines (Chambers 1978) and Rishworth Moor, near Soyland (Bartley 1975). That altitude has an important role to play in the denseness of the woodland cover is shown by the lower values of 30-50%

TP in Upper Teesdale, at 893 m OD (Turner *et al* 1973) compared with the average figure of 60-70% TP in lowland sites (eg Neasham Fen, in lowland Durham (Bartley *et al* 1976), and Crose Mere, Shropshire (Beales 1980)).

During the immediate pre-Elm Decline period, the extent of the mire increases enormously, and a study of the four diagrams at Soyland can now give some information of the geographical position of the woodland on Soyland Moor. At this period at SMC the values drop from an average of 45-55% TP to consistently below 50%. At SMD, this is not seen until the very end of zone 5, and indeed, the AP values increase slightly through the early part of the zone. At SMA and SMB, the arboreal values are consistently below 50% TP. This could be simply explained by the greater abundance of trees on the northerly side of the bog, but Tauber's model demands a more complex explanation. SMC and SMD, now in the middle of an extensive bog, should be representing more regional landscapes than SMB and, in particular, SMA, which are much nearer the edges. The lower arboreal values near the edges of the bog indicate that it is local woodland that had decreased. This is, of course, partly due to the spread of peat itself, but, as a corollary to this, conditions for tree growth may have been less favourable across the plateau top. The higher values, particularly at SMD, represent the more regional pollen rain, probably from up to several kilometres away, and as such reflect the dense woodland from the sheltered valleys at lower altitudes where the increasing wetness was not yet producing problems with waterlogging, and indeed may even be enhancing the conditions for tree growth, alder in particular.

Some researchers add *Corylus* values into the arboreal pollen sum (eg Turner and Hodgson 1979, 1981), due to its propensity to form a small tree in open situations. If this is applied to most pollen diagrams, then the picture of open woodland in the uplands is altered radically. An average figure of 70-85% AP is then applicable to both the uplands and the lowlands, indicating closed woodland throughout Britain (Tinsley and Smith 1974, Litchi-Federovich and Ritchie 1965). However, these figures do not take into account the over-representation of hazel pollen in the uplands, which occurs when it is growing in the open rather than as an understorey shrub (Godwin 1975a). If corrected, the arboreal values in the uplands decline to 60-65% of TP.

Thus, woodland cover in the uplands may well be considerably less dense than in the lowlands. The composition differs mainly in that hazel played a constantly important role in the uplands, but all the tree species of the lowlands were present. At Soyland, however, the long continuity of pine and the stable birch communities underline differences between the uplands (at least in the central Pennines) and lowlands.

The above discussion has centred on the denseness of forest cover across the landscape at different altitudes rather than the possibility that lower arboreal values may indicate the position of actual tree lines. Tauber's model cannot incorporate vertical as well as horizontal parameters, and canopy and rainout components are not differentiated in this way. *In situ* macro evidence has to be used to look at tree-lines. Tree stumps and branches are common in and below the peat in the central Pennines. The lack of any macro remains in the mire itself reflects the probable continuously waterlogged surface from its inception, which would inhibit any tree invasion onto its surface. Evidence from the surrounding areas confirms the pollen evidence that Soyland at 389 m OD was well below the tree line which occurred at c.500 m (Tallis and Switsur 1983). Pine, birch, willow and even alder are found above the level of Soyland. The lack of other species does not necessarily mean that they did not grow at this height, but merely that they are less likely to grow in areas where they will be incorporated in the peat (Tallis and Switsur 1983). The pollen evidence, at SMA particularly, suggests that even lime, the most thermophilous of Britain's trees (Grieg 1982), managed to grow at 389 m for a short period. Together the evidence suggests the altitude at Soyland is not great enough to stop the invasion of any of the tree species growing in lowland Britain at this period.

To summarise, the interpretation of the maximum arboreal values of 45-55% TP as 'open woodland' may well be an over-simplification. The inclusion of corrected hazel values brings the figures a little closer to that of closed canopy woodland, but the exposure factor may affect the density of cover with more open woodland on the fully exposed areas and dense stands developing in more sheltered areas such as in the lee of Blackstone Edge. The spread of blanket peat diminished the area available for tree growth but the increasing wetness did not seem to perceptibly affect tree growth itself and small stands of deciduous forest remained and even increased a little. The rest of the landscape remained dominated by birch, hazel and heathland. The possible role of man in this process is discussed in the next chapter.

The distribution and extent of individual tree species

In this section, the position of each species, as far as can be ascertained, is discussed. Some of the information has been discussed zone by zone in the previous chapter. Here it is summarised and brought together.

Birch: the *Betula* values are probably representative of *B. pubescens*, the silver birch (Godwin 1975a).

The problems of the initial over-representation and subsequent under-representation of birch have been discussed in Chapter 4, and the interpretation of the values must remain rather difficult. Birch macro remains supplement the pollen values and indicate that birch was a major component of the woodland up to 520 m in the central Pennines (Tallis and Switsur 1983, Tallis 1975), and can be found even in very exposed situations. At Pule Hill, an isolated and exposed hill to the south-east of Soyland, birch remains were found at 425 m (Woodhead and Erdtman 1926). In the course of fieldwork for this project, birch was often found in peat hags in this area (identified by J Hillam of the Dendrochronology Laboratory at Sheffield University). The highest site was at Pudding Real Moss on Wessenden Moor (SE 077078), at 480 m OD. The ready availability of birch is underlined by its common appearance in Mesolithic hearths (eg Woodhead and Erdtman 1926). While these macro remains may well give a biased picture of the position of birch within the woodland, it does corroborate the evidence of its low pollen productivity in some upland areas (Tinsley 1972).

The relatively steady nature of the birch pollen profile after its early postglacial decline suggests it had established a population within a particular habitat, most likely on the damper areas of the plateau. Once established, none of the later invading species appear to have successfully competed with birch. The slight and gradual decline through time is most likely to be a percentage effect due to the increasing number of trees in the immediate area, which would mask out some of the Betula pollen. The small peaks of Betula values which occur through the diagrams may represent the invasion of birch into abandoned clearings which were dominated by Calluna heath. Corylus subsequently may have taken over these areas. Thus birch may have a seral role in the regeneration of clearances, similar to that postulated by Dimbleby (1952a). Robinson's Moss in Longdendale shows a very similar birch profile (Tallis unpublished), suggesting that the position of birch is uniform above 300 m in this area.

Pine: Pinus sylvestris (Scots pine) is the only native Pinus species in Britain (Godwin 1975a).

The extended pine maximum and the position of pine stands on well-drained rises around the mire have been discussed in the previous chapter. This showed that pine held an important position for much of the pre-Elm Decline period. Birch was once thought to constitute the major macro remain in the central Pennine peat, but it has now been established that pine and willow are marginally more common than birch (Tallis and Switsur 1983). A large area of pine stumps, unfortunately undated, has been found at Sunderland Pastures, just 4 km to the north of Soyland at 365 m (J H Tallis pers. comm.), confirming that pine could and did grow in pure stands. The decline of pine to such low levels in the immediate pre-Elm Decline period suggests it disappeared from the plateau top altogether. The slightly higher remnant pine values at SMD compared to the other three sites could reflect the position of any remaining trees on the north-west side of the bog, or may be reflecting the more regional pollen rain where some trees may remain in the valleys. Its total decline has to be seen in terms of its replacement by oak, which would prefer the better drained areas to the increasingly wet plateau top.

Elm: The Ulmus profile is probably representative of U. glabra, the wych elm (Godwin 1975a).

After its initial rise, elm values are relatively stable through time and space on Soyland Moor. There are temporary decreases in values at 2915 ± 50 bc at SMC, at 3870 ± 95 bc at SMD and during the long 'Potentilla-type peaks' of the pine maximum at both SMC and SMD. These declines are most probably associated with clearances and will be discussed later, as will the Elm Decline itself. The average value of 4% TP at Soyland is similar to those at Robinson's Moss (Tallis unpublished), and to Rishworth Moor (Bartley 1975). However, some similarly situated sites show much lower elm, making the position of the Elm Decline very difficult to place, for example, Leash Fen in Derbyshire (Hicks 1972) and Fountain's Earth in Nidderdale (Tinsley 1975). The values of elm pollen of 4% TP may suggest that elm was well established right to the tops of the nearby valleys but did not successfully establish on the plateau top (Tinsley and Smith 1974). However, the temporary declines in elm values during clearance phases may indicate small populations on areas of slightly richer soils on the plateau top itself. Exposure and poor soils are the standard explanation for the low populations of elm in the higher areas of Britain. However, analysis by Turner and Hodgson (1979) in the northern Pennines, while they do correlate high elm values with base-rich bedrock types, also correlate them with very high altitudes, usually over 500 m. Thus, exposure does not seem to be an important factor in limiting the spread of elm.

Oak: Quercus pollen includes both Q. petraea and Q. robur. Q. petraea most probably is the most dominant oak species in the northern uplands (Godwin 1975a).

Oak was present on the plateau top from 6100 ± 50 bc, accepting a figure of 5% TP to indicate local presence (Tinsley and Smith 1974). However, it did not expand to dominate the woodland until two millennia later at 4160 ± 40 bc. As previously discussed, this delay was probably due to competition from pine. The onset of wetter conditions coincided with, or may even have been the cause of, the full expansion of oak at the expense of pine. During the early period it is possible that oak formed a light woodland with hazel. After 4160 ± 40 bc, oak values increase substantially. The concomitant drop in hazel values may suggest hazel now formed only an understorey in denser oak woodland, rather than an open woodland where it was co-dominant with oak. If, as most researchers now agree (eg Janssen 1967, Tinsley and Smith 1974), oak is not over-represented in the pollen rain, then oak had become the major woodland component, not excluding the over-represented hazel. In the immediate pre-Elm Decline period, oak values increase slightly from SMA through to SMD. This could be taken to confirm that oak did supplant pine on the north-west side of the bog, where pine showed more local presence in the earlier period. However, the oak values at SMD, while a little higher, are not particularly erratic (and therefore indicative of local presence) and this phenomenon may again be explained by pollen representation in large mires (Janssen 1966), implying that oak was more important in the surrounding valleys than on the plateau top. The loss of oak during SMC 5, which is slightly echoed at SMD 5, is related to the clearance phase at 2915 ± 50 bc and will be discussed later.

Alder: Alnus glutinosa (alder) is the only native species (Godwin 1975a).

The slow and gradual rise of alder is not necessarily connected to the prolonged pine maximum, as is the slow rise of oak, although the increased local pine pollen input may mask greater amounts of alder regionally. At Soyland, alder values benefit from the increasing wetness after 5000 bc, but it is representative of an established species slightly extending its range rather than a sudden population explosion from isolated stands, as is usual in British pollen diagrams.

At the SMC and SMD 4/5 boundaries, dated to 4160 ± 40 and 3870 ± 95 bc respectively, alder values show a substantial increase, associated with a sudden peak in Sphagnum. This may indicate a further period of wetness which was more critical than the previous one, allowing alder to substantially increase its population. The alder curve at Robinson's Moss follows a similar pattern echoing echoes this increase, although it is not as clear a phenomenon. The increase at Soyland is not maintained right through to the Elm Decline and alder suffered with the other trees during the reduced arboreal values of the immediate pre-Elm Decline period. This may be merely a reflection of the extended area of bog, but the sharp decrease in values suggests actual diminution of alder's population in the immediate vicinity. As conditions during the wet Atlantic period are favourable for alder, clearance by man may be the explanation of the loss.

The actual position of alder in the landscape is difficult to assess. Its high values in the later period, where it is second only to oak, indicate that it was present on the plateau top in some quantity, and must have been part of the woodland, as well as along the numerous streams that would occur across the millstone grit. Godwin (1975a) states that alder is unlikely to be a component in the deciduous forest at high altitudes, but macro remains are found up to 425 m in the central Pennines (Tallis and Switsur 1983). At 389 m, Soyland Moor is well within its altitudinal limit. A complicating factor is alder pollen representation. Firstly, Tinsley and Smith (1974) found that modern alder pollen has erratic production and can show high values where it is only a minor component of the vegetation. Secondly, alder, as a catkin bearer, can over-produce pollen generally (Smith 1972). The smooth curves at Soyland suggest that erratic pollen production is not a problem, or perhaps that it is smoothed by distance, and certainly do not reflect alder growing on the edges of the mire. If alder is not situated anywhere near the mire, then its high values must be the result of gross over-representation of alder carr communities in the marshy valley bottom.

<u>Lime</u>: Both <u>Tilia platyphyllos</u> and <u>T. cordata</u> are included in the <u>Tilia</u> pollen type. <u>T. cordata,</u> less demanding in its ecological requirements than <u>T. platyphyllos</u> (Moore 1977a), is most likely to be the species represented here (Pigott and Huntley 1980).

The evidence for lime in pollen diagrams in the north of England seems to indicate that it was generally not an important element within the woodland. However, some northern sites, both upland and lowland, do have pollen values which suggest that lime may have been an important element of the woodland in some circumscribed areas (eg areas of the North York Moors (Simmons 1969), and areas of the Durham lowlands (Bartley 1966)). The <u>Tilia</u> values from Soyland Moor and Robinson's Moss can be interpreted as the local presence of lime on the plateau top, probably in pockets of more base-rich soils. The extremely poor pollen dispersal of lime may then be the reason for its low values at many northern sites. The high values at SMA 2 may indicate a small area of woodland dominated by lime, probably extending from the top of the River Ryburn valley under the sheltered lee of Blackstone Edge. The sudden decrease in <u>Tilia</u> values at SMA 2 represents the decline of its dominance in this small area, but it remained a component of the woodland across the plateau.

<u>Ash</u>: <u>Fraxinus excelsior</u> is the only native species of ash in Britain (Godwin 1975a).

The under-representation of this species and the tiny amounts recorded at Soyland make more than general statements on the position of ash impossible. It is a light demanding tree which can tolerate exposure and soil moisture (Clapham, Tutin and Warburg 1962), and as such should maintain a reasonable population in the upland forests. However, its preference for fertile soils may keep its population small, particularly as its late establishment in Britain means that it had no established niche from which to spread. The erratic nature of the ash profiles is likely to reflect the nature of the ash population which took advantage of openings in the woodland to quickly spread, but could be just as quickly shaded out again.

Sorbus: Sorbus aucuparia, the rowan, can be differentiated as a separate pollen type. Sorbus undiff. can include most of the many different native species.

Sorbus, like Fraxinus, is a very under-represented genus (Tinsley and Smith 1974), and its sporadic occurrence at Soyland is probably indicative of a persistent population (Godwin 1975a). If this is the case, it confirms the open nature of the woodland cover (Mamakova 1968). In the early stages, Sorbus is more important at SMD than SMC. This in part reflects the greater age of SMD, as the majority of the records here are from the period when peat had not yet formed at SMC. However, throughout the early period, Sorbus appears to have been more important on the north-west side of the mire. After pine becomes dominant, records of Sorbus are sparser at SMD, reflecting the increased forest cover of pine and then oak seen at SMD. By SMD 5, Sorbus has disappeared entirely. Its occasional appearance at SMA, B and C indicates it was still present in the locality.

Hazel: Corylus avellana is the only native British species (Godwin 1975a).

The problems of interpreting hazel values are in part due to our lack of substantial autecological knowledge of hazel, and in part due to its variety of response in pollen and population terms to changing circumstances, and also its close association with man in the Mesolithic period (eg Jacobi 1978b). Further, modern pollen studies show it has an erratic pollen production even within one type of habitat. Tinsley and Smith (1974) state that only figures of over 20% TP can be safely taken to indicate local presence. Despite these problems, hazel seems to show a reasonably standard response during the postglacial, altitude being the only major factor for change, and Soyland Moor is no exception. Its values rarely drop below 15% TP, indicating a constant local presence on the plateau top. Its overwhelming dominance in the early stages of the postglacial shows its common seral role in woodland development. Throughout a gradual decline can be seen as other trees invaded. How much of its decline is due to reduced pollen production and how much due to its actual replacement is difficult to judge.

The possibility of man having a decisive role to play in the spread and maintenance of hazel by burning and coppicing is discussed in the next chapter. The more erratic nature of the hazel profile in SMC 5 may indicate the increased role of man on a small scale, as it has less correlation with the corresponding zones in SMA, B and D than previously, indicating local differences where before they were similar.

The Plateau Soils

Today, the Soyland Moor area is peat covered, and the surrounding soils are very acidic, supporting ericaceous communities. The evidence for the state of soils on the plateau during the postglacial is contradictory. Generally, it is agreed that soils have matured after the glacial period by the end of the Boreal period (eg Dimbleby 1965), but there is less agreement on when leaching and podsolisation occurred. Dimbleby (1952a, 1962a) states that full podsolisation is rare before the Bronze Age. This conclusion is based on direct evidence of buried soils below archaeological monuments in the North York Moors. However, different lines of evidence suggest podsolisation occurred earlier in the Pennines. Conway (1954) states that mor humus replaces mull during the mid-Atlantic on pollen and stratigraphic evidence. Tallis (pers. comm.) states that soils below peat which formed after the mid-Atlantic (c. 4000 bc) show podsolisation, whereas soils under earlier peats do not.

The lack of recent glaciation, the largely unknown effects of periglacial activity and the base-poor parent material in this area make it difficult to assess the nutrient status of the pre-4000 bc soils. Periglacial activity would have increased the available nutrients (Taylor 1965), as can be seen in the initial _Salix-Betula_ carr in SMD 1, with its rich fen affinities and calcicole flora. The early loss of this rich flora and the high values of ericaceous species throughout the diagrams suggest that some parts of the plateau at least were leached and subsequently podsolised at an early stage in the postglacial, unless all the ericaceous species were confined to the bog surface, which is improbable. However, many species present indicate that some localised areas remained relatively base-rich. The high hazel values could not be maintained on very acidic soils (P D Moore pers. comm.). The long pine maximum, on the other hand, could have an initiated raw humus development with a tendency to leaching, especially if pine existed in pure stands (Taylor 1965), but the succeeding oak woodland may have reversed this trend by improving the pH status, and thus bringing back a brown earth soil (Smith 1972) to some parts of the plateau. _Sambucus,_ which is associated with nitrogen-rich soils (Godwin 1975a) is important only in the pre-4000 bc period at SMD, although it does occur once in SMA, post-4000 bc.

The evidence of the entomophilous herbs with their low pollen productivity and poor dispersal, may more surely indicate the status of the soils in the immediate neighbourhood. Those herbs which are generally confined to relatively base-rich soils are not common, and do show a slight decrease after 4000 bc, although they do not disappear altogether. This indicates that, while the soils were never base-rich, they remained less acidic than today. Thus _Poterium sanguisorba, Saxifraga granulata_ and _Geum_ were present before 4000 bc only, but _Urtica dioica_ remained common throughout, as did _Filipendula_ which is a reliable indicator of non-leached soils (Moore and Bellamy 1974), although it could be present on stream banks only. The only occurrence of _Malva_ is post-4000 bc, in SMA. The occurrence of such species as _Urtica_ and _Sambucus_ may be related to the temporary enrichment of the soils induced by burning, but _Malva,_ for example, is independent of man's activities. Those species which indicate a lack of nutrients are mainly confined to the bog which has been ombrogenous from the early stages. Cf *Jasione montana* in SMD 4, however, indicates sandy, lime-free soils.

Thus the pollen evidence does not give decisive information of the status of the local soils during the postglacial. They cannot have been very rich at any time after the initial post-glacial period, but do not show the extreme impoverishment of such upland soils today. Their ability to maintain deciduous woodland in itself will have helped to maintain soil nutrients and effectively slow down or stop leaching. The woodland ecosystem would appear to be fragile, easily reverting to *Calluna* heathland, under which podsolisation would occur (Handley 1954). However, although *Calluna* remained important on some areas of the plateau throughout the early postglacial, the heathland was often invaded by *Betula* and subsequently by *Corylus*. For this to occur, the soil pH must have remained above the critical level of 4.5 (Miles 1978).

Summary

This chapter has followed the development of Soyland Moor from its initial stages of willow-birch carr with its rich fen affinities through the development of the topogenous mire, to the spread of the blanket bog over much of the present area of the moor. The increasing area of woodland surrounding the mire was discussed in terms of its composition, geographical extent, where this was possible, and the maximum area of woodland as compared to other areas of Britain. This latter is confused by the difficulty of interpreting the hazel profiles, but in general, the woodland was open on the plateau top, with herbaceous species and dwarf shrubs maintaining much larger communities than in the lowland sites and, indeed, some upland sites. Pine was the dominant tree species for a much longer period than at most British sites, and woodland cover reached its maximum extent during this period, which may indicate that local conditions were near the threshold inhibiting deciduous woodland growth. However, the establishment of lime in the later period suggests that this was not the case. Man may well have played a decisive role in the density of the woodland cover. The base status of the soils on the plateau was initially, in the early postglacial, quite high, although early leaching took place. However, throughout the pre Elm-Decline period the soils remained less impoverished than today.

CHAPTER 6

THE EFFECT OF MESOLITHIC MAN ON THE VEGETATION OF SOYLAND MOOR

The impact of Mesolithic man on the early postglacial environment was first studied systematically by Dimbleby (1961a) on the North York Moors. Gradually, the idea that Mesolithic peoples were hunter-gatherers of the simplest type, existing in a random and precarious fashion without disturbing his environment (eg Iversen 1949), was discarded. Today, most palynologists working in the uplands of Britain can point to clearance phases prior to the Elm Decline (see, for example, the articles by I G Simmons and R L Jones). Often, however, comment is confined to the AP:NAP ratio and a list of associated ruderals (eg Tinsley (1975) at Fountain's Earth, Nidderdale). Relatively little attention is paid to the mechanisms involved, or the possible long term effects of these seemingly ephemeral clearance phases on the vegetation. The possibility that they may not be anthropogenic in origin is not often explored. Post-Elm Decline clearances still attract more interest, despite their essential similarity to earlier phases of forest destruction (Smith 1970), and much information on the Mesolithic is missed due to large sampling intervals (Smith 1982).

Most of the systematic work on Mesolithic clearances has been undertaken in the area they were first discovered, in the North York Moors, under the aegis of Professor I G Simmons and, especially in terms of resultant soil degeneration, by Professor G W Dimbleby. The central Pennines can boast of only one systematic study (Jacobi *et al* 1976), but this clearly pointed out the effects of what appeared to be a long term and widespread policy of clearance in the area.

The first part of this chapter looks at the hypothesis of Mesolithic burning which is now widely accepted (Simmons 1969, Mellars 1975, 1976b). The second half of the chapter looks at the evidence at Soyland Moor within the framework of this hypothesis.

The Hypothesis of Mesolithic Clearance

The use of fire

As evidence for Mesolithic burning accumulated, and the idea became acceptable, the theoretical and cultural aspects were studied, notably by Mellars (1975, 1976b) and Simmons (1969, 1975a). It was pointed out that, in ethnographic terms, the use of fire by primitive peoples was, and is, almost universal (Jacobi *et al* 1976). Thus on *a priori* grounds, the use of fire in the Mesolithic (and before) should be considered as a possibility. The modern literature on the effects of fire on plants, animals and the soil is prodigious, but has to be used carefully. Ralston and Hatchell (1971:68) state the problem: "Almost any position on fire effects (ie good, bad or indifferent) can be documented by reports of reputable researchers. Precise conditions need to be known." The precise conditions of the postglacial are not known, which means detailed study of the effects of burning on the environment is difficult. Also, much of the research has been undertaken in the USA in coniferous forests whose history is inextricably bound up with the regular occurrence of natural fires

(Komarek 1966), and as such is not applicable to the European temperate forests. However, enough information is available to make some generalisations on the use of burning. Controlled fires, usually of low intensity, can result in a temporary increase of available nutrients in the soil via ash, and increase the amount of sunlight reaching the forest floor by clearing woody or dead undergrowth (in the case of open woodland) or by the opening of the canopy (in closed woodland) (Mellars 1976b). The direct result is to stimulate new and luxuriant growth of understorey herbs and shrubs and suckering in some trees (Mellars 1975).

The objectives

The increasingly wooded environment of the postglacial would have imposed economic stresses on Mesolithic peoples (Mellars 1974). Edible vegetation was reduced as the tree canopy became denser and hunting became more difficult as the large herds of the late glacial period disappeared (Simmons 1969). Reindeer and elk were replaced by red deer, wild pig and aurochs, smaller animals which congregated in smaller herds or were solitary, living in an environment with an increasing amount of cover (Komarek 1967, Simmons 1975a). In this situation, burning of the forest undergrowth (and occasionally the canopy) would have had a number of advantages. In ethnographical terms, a major advantage is the improvement of herbivore browse (Mellars 1967b). Herbivores are attracted to burnt areas and the density of animals within a given area can increase by 400-900% after burning. The body weight of individual animals increases due to better quality browse. Total carrying capacity can also be improved as good summer browse advantageously affects the ability of herbivores to survive the critical winter months, and reproductive capacity is often improved (Jacobi *et al* 1976). By burning selected areas, hunters could predict the movement of herds: human population density is controlled not so much by the absolute amount of food available but by the predictability of obtaining it (Mellars 1976b).

In temperate regions, vegetable resources are as important, if not paramount, to man's survival as the animal resources (Bradley 1978). Burning can increase both the quantity and quality of these resources (Mellars 1976b), particularly of edible open ground herbs (eg Chenopodiaceae, *Urtica dioica* and Leguminosae). The yield of berry and nut producing species is increased (Wright and Heinselman 1973) and, in the Mesolithic context, the prevalence of *Corylus* is of particular importance. Again, burning may help in predicting the size of the harvest, as it can stabilise the notoriously unpredictable yield of wild nuts and berries (Komarek 1969), and will also facilitate harvesting (Mellars 1976b). Other possible uses of fire are clearance of dead or rank undergrowth to facilitate mobility, important during seasonal movements, and for hunting. However, the loss of too much cover for forest dwelling animals can cause them to leave the area, so a balance must be maintained (Simmons 1975a).

The mechanism

The study of prehistory is concerned with long term processes and the reasons and mechanics of culture change, rather than individual incidents. Thus, while individual clearance phases picked out from the pollen record are of interest, they do little to further our understanding of Mesolithic economy and society. If the hypothesis of Mesolithic burning as stated by Simmons (1969) and Mellars (1976b) is of long-term importance within the economy, then evidence to test it must be in the form of repeated phases of clearance activity throughout the period. Only then can the study of the Mesolithic economy and the change to the Neolithic incorporate the more sophisticated model demanded by the hypothesis. Three subsidiary lines of evidence are also of use in testing the more specific aspects of the hypothesis:

1. The hypothesis indicates that burning would be carried out during the late spring and early summer, as most researchers into modern fire effects agree that burning at this season produces the best results (eg Komarek 1967). Irwin (1975) points out that herbivores will use burned areas in the summer, unburnt in winter. Also, burning may well be carried out mainly in areas with acid and sandy soils; acid soils derive most benefit from burning (eg Lutz 1956), and thus would be burnt before neutral or basic soils; sandy soils carry vegetation which is more likely to burn (Jones 1976b).

2. The advantageous effects of burning are temporary. The time span is of course dependent on many factors such as the vegetation and soil type, the length of the growing season and the intensity of the burn (Mellars 1976b), but generally, in deciduous forests, maximum nutrients are available in the first or second season after the burn, and the maximum vegetative growth in three to four years. After c. 15-20 years the effects of a controlled burn should have disappeared (Mellars 1976b). Thus repeated clearances would have been necessary if fire was an integral part of the Mesolithic economic strategy. Clearances would have to be maintained at various stages of development over the summer range (Jacobi et al 1976, Simmons 1975b).

3. The size of the area cleared would be relatively small, if, as expected, most clearances were made for the use of herbivores (Mellars 1976b). A figure of roughly 200-250 m maximum diameter is given by Mellars (1976b). If much larger, herbivores will avoid the centre as being too far from cover, but, if very small, the very proximity of cover would make culling too difficult to allow for predictable supplies. It is likely too that small areas only would be burnt for vegetable resources as they would have to be guarded from depredations by herbivores. Placing such clearances near to the habitation sites would help in keeping the herds away. Larger areas could be fired however if mobility was the paramount objective. Clearance could rarely be multi-purpose, as herbivores would avoid areas too often frequented by man.

The evidence for burning at Soyland Moor and other areas, will be discussed within this framework.

Problems in Studying the Evidence for Clearance in Pollen Diagrams

Most Mesolithic sites in the uplands comprise flint tools and waste only, with the occasional post-hole, charcoal and charred hazel nut. The acidic soils do not preserve bone. Thus, evidence of economic strategies has to be looked for elsewhere. The widespread occurrence of peat deposits in the uplands has made palynology the most productive method for studying the possible use of fire by Mesolithic man. However, the method has to be used with discrimination, as the evidence is indirect and subject to misinterpretation (Edwards 1982, Rowley-Conwy 1982). Much of the work in interpreting clearance phases has been done for the post-Elm Decline period (eg Turner 1964, 1965), but it can usually be applied back to the Mesolithic (Smith 1970).

The first major problem is finding deposits which will yield evidence of clearance. Up to a point, this is a hit and miss affair: one can only be certain after all the hard work has been done. However, the second problem area, concerning assumptions about the evidence, can be corrected and more sophisticated models evolved to help analyse any available evidence.

In the first case, it is possible to select sampling strategies which will maximise the chances of finding evidence. Standard sampling strategies, as discussed in Chapter 3, may often mask the evidence for clearance. The major problem here is the distance of the pollen source from the pollen deposit, given the exponential fall-off of pollen representation (Edwards 1982). Many pollen studies are from cores taken in the middle of large mires or lakes, where local pollen patterns are masked out by the much greater amounts of regional pollen rain (Tauber 1965, Janssen 1967). If specifically looking for evidence of clearances, then small deposits are necessary, within an area where the archaeological evidence indicates that clearance may be a likely strategy (Rowley-Conwy 1982). The use of three-dimensional pollen diagrams (eg Turner 1974, Jacobi et al 1976) can also help in pinpointing clearances on a larger scale.

Very short clearance phases may not appear in diagrams, especially if total regeneration of the woodland occurs afterwards (Edwards 1979a). This can be partly countered by close sampling (Smith 1982), and by choosing deposits, where possible, with a fast rate of growth, giving greater resolution of the pollen record. However, even with an optimal sampling strategy, one sampling interval may contain the evidence of burning, the use of the clearance and its abandonment, and regeneration. The mixture of these phases may well prove indistinguishable. The clearance of a small area of ground, as may well be the case if the strategy was to encourage herbivores, may not appear in diagrams even given close sampling, and even if the deposit is situated close to the burnt area. Characteristic ruderals may not travel far through the surrounding woodland (Turner 1964, Smith 1982). This however, can be a useful characteristic in that the appearance of a wide range of ruderals is more likely to represent local clearance only, and the changes in the AP:NAP ratios can be used to discuss the size and type of the clearance.

Thus, conventional sampling strategies will tend to grossly under-represent evidence for clearance activity, and even an optimal sampling strategy depends to a certain extent on chance.

The assumptions made by palynologists regarding pollen evidence can be simplistic; ie a clearance episode is short and appears in the diagrams as a reduction in arboreal pollen values associated with an increase in the variety and abundance of ruderals and some light-demanding shrubs and trees such as hazel and ash. However, this is by no means the whole story, and occasionally the wrong story altogether. There are a number of circumstances in which AP values will remain steady or even increase during a clearance phase:

1. Flowering may be greater due to extra light (Vuorela 1970), producing extra quantities of AP.

2. Clearances can allow better dispersal of the surrounding AP (Smith 1982).

3. The tree canopy may not be affected by low intensity burning (Mellars 1976b).

4. The burning may take place in areas of scrub (particularly hazel) rather than woodland.

5. Repeated clearance is necessary to kill mature trees and stop immediate seedling regeneration.

6. During a period of tree invasion, generally rising AP values may mask any local loss.

On the other hand, a loss of AP values may be the result of natural processes, such as local windthrow or a lightning fire, or to the vagaries of pollen production, such as the replacement of oak by lime: the number of trees may not diminish, but the lower pollen production of lime would reduce the AP values.

The use of herbs and shrub species as indicators of man's activities can also be misleading. Apart from cereals and known imported weeds, no species are directly associated with man. It is the way man manages the environment which brings about the association, and it is possible that these associations have changed since the early postglacial. Ruderals were very common in the late glacial period, but their abundance dramatically declined with the invasion of trees during the early postglacial. Only after the introduction of farming on a large scale do ruderals regain an important role in pollen diagrams (Godwin 1949). This late association with man's activities has been transferred to the appearance of ruderals in the Boreal and Atlantic now that the idea of Mesolithic clearance has been accepted. <u>Plantago lanceolata</u> is a good example of this: for a long time, it was thought not to occur in British pollen diagrams until after the Elm Decline, where it appeared to be always associated with man's activities. When it was subsequently found in pre-Elm Decline deposits, its association with man was assumed (eg Smith 1970). However, there are a number of other possible explanations for the presence of ruderals in pollen diagrams. Firstly, for the sudden increase in ruderals to occur in the post-Elm Decline period, refugia must have been present for these species throughout the early part of the postglacial. The more open woodland of the uplands were the most likely sites for refugia. Secondly, both animals and man make trackways which provide natural habitats for ruderals (Thomas 1960, Proudfoot 1971). Trackways could then be mistaken for small clearances if they occurred close to the pollen deposits. Thirdly, unstable boundary zones are found on the coast, in river valleys and any area which is the boundary of two highly contrasting habitats (Groenman-van Waateringe 1968). Such zones are often dominated by herb-rich open communities and would attract herbivores which would trample the ground, creating habitats for ruderals.

The high values for hazel in the postglacial period are often interpreted as the result of Mesolithic clearance, due to the ability of hazel to quickly invade after burning (eg Rawitscher 1945, Perring 1965, Jacobi 1978b). Smith (1970) points out that hazel nuts are often found on Mesolithic sites and may well represent a staple resource, so that burning could be a deliberate policy to increase the amount of hazel in a nut-bearing condition. However, the initial high values of hazel in the postglacial are more readily explained by refugia close to, or in Britain in the late glacial from which it would easily invade into the largely open landscape of Zone V (Deacon 1974), rather than by widespread burning. Its continued importance may be explained merely as inertia within the existing vegetation (Smith 1965), helped by the existence of only open woodland in the uplands. Of course, it is possible that a combination of early migration and man's activities are the explanation for the high postglacial hazel values. Indeed, the existence of 'only' open woodland itself could be a result of burning. However, the high values for hazel over much of upland Britain cannot be used as an anthropogenic indicator until evidence for widespread and prolonged clearance is documented. It is also possible that hazel was not actually encouraged by man. The evidence for hazelnuts in Mesolithic hearths (eg Barnes 1982) shows merely that they were used as food, but their importance in their diet is not known.

The appearance of most other shrubs apart from *Salix* is also often used as evidence for clearance (eg *Ulex, Prunus* and *Sambucus*), especially as many of these do/not occur in any quantity until after the Elm Decline (Godwin 1975a). The explanation for their appearance before the Elm Decline, as with the ruderals, may merely lie in the existence of upland refugia. *Fraxinus* pollen also is often seen as first appearing, or increasing, with clearance (Godwin 1975a), as is *Fagus* (Simmons 1969). The unstated assumption is that these tree species would not have been established were it not for man-made clearances.

Despite this far-reaching assumption, and despite statements that Mesolithic man may be responsible for the countrywide hazel maxima (eg Perring 1965) and for peat initiation in the highlands (eg Moore 1975a), it is still assumed by many researchers that Mesolithic clearances were ephemeral, with little or no permanent effect on the vegetation (eg Simmons and Cundill 1974a). While this may be the case, neither suggestion is valid until demonstrated and it is possible that evidence for large scale clearance is overlooked, because of assumptions about the economic sophistication of Mesolithic man. (If longer and/or larger clearance phases are picked out, there can be a problem in deciding whether the evidence depicts rotation of small scale clearances or widespread group activity (Edwards 1979b).)

Thus, dogmatic and simplistic assumptions on the appearance of clearance evidence in pollen diagrams are misleading: often a case of missing the wood (or lack of it) for the trees. Statements of AP:NAP ratios and lists of ruderals which so often comprise the full analysis of clearance evidence gain credence merely by reiteration. For the evidence to be interpreted as anthropogenic a full range of interlocking evidence is necessary, plus some understanding of the economic strategy it is supposed to represent. Some archaeological backing would also be helpful, although this is not always possible, and is often confined to assertions that a densely occupied area is, on *a priori* grounds, more likely to produce evidence for clearances than areas which were scantily inhabited.

The Interpretation of Clearance Phases in Pollen Diagrams in Terms of Management

In an attempt to produce models a little more sophisticated, the following proposals are based on differing techniques of producing clearings: thus, fire, felling, ring-barking and grazing pressure are also included.

Clearance for animal resources

The clearance of forest or scrub for grazing purposes was probably the most widespread management technique in the Mesolithic (Jacobi *et al* 1976). This can take a number of different forms, differing mainly in intensity. Forest grazing is an extensive technique, and consists of initial low intensity burning, leaving the trees undamaged (Behre 1981). Grazing pressure then dictates the successive vegetation. If light, then shrubs, particularly *Corylus*, will rapidly invade. Light grazing pressure will occur when the management technique is confined to attracting the animals by burning. Clearances at different stages of re-colonisation would be maintained in a mosaic across the landscape. This would appear in pollen diagrams in two ways. If all the clearances in differing stages of succession occur near to the pollen deposit then the following features would occur:

1. No loss of AP values.

2. Possibly an initial loss of *Corylus* values indicating the burning itself, but most likely an increase will be registered immediately and for the duration of the use of the area. The *Corylus* profile would not be smooth, showing peaks and troughs representing the differing stages of re-colonisation of the numerous clearings.

3. A small increase in *Gramineae* values.

4. An increased occurrence of open ground indicators such as *Pteridium* and Umbelliferae (Behre 1981). However, highly characteristic herbs of grazing, such as *Jasione* (Behre 1981), may not occur, as these represent areas of grazing on grass swards which would not have time to develop in these ephemeral clearances.

Secondly, if the centre of clearance was located away from the pollen deposit, then only the outlying clearings may be represented. This would appear in the pollen spectra as very short, ephemeral phases lasting only for one or two levels. They would have similar pollen characteristics, but the *Corylus* profiles would be smooth.

Heavy grazing pressure represents a more intensive management technique with some element of confinement by fencing or driving (otherwise the herds would leave the area to find better browse). Regeneration of trees would be halted as seedlings would be grazed (Pearsall 1968), and Corylus and other shrubs would be reduced in number or even eradicated (Dimbleby 1978). Eventually a grass and ruderal sward would develop under the remaining mature trees. This type of clearance would appear in pollen diagrams in the following way:

1. No loss of AP values, unless the phase occurred over a long period of time, when lack of regeneration would reduce the number of trees. Thus AP values may decline towards the end of the clearance.

2. Corylus values would decline, unless a very small area was maintained in this way, when increased flowering around the edges of the clearance could result in maintained values.

3. Gramineae values would show the major increase soon after the beginning of the clearance.

4. The more characteristic pasture ruderals would appear: Melampyrum, Plantago lanceolata, P. major/media, Jasione (Behre 1981), and Potentilla erecta (Mamakowa 1968).

5. If the pasture was wet, then Succisa and Ranunculus acris type may also be common (Behre 1981).

Another form of management would be the creation of open low scrub or meadows. Here the trees themselves would be burnt or cut down to give man more control over the herds due to better visibility. The creation of open scrub would be a less intensive form of management and would be represented in diagrams in a similar way to the mosaic of small clearances discussed above (except that AP values would decline if the areas cleared were extensive). The creation of meadows would require some form of confinement, as the development of a grass sward would not necessarily attract grazing animals. Heybroek (1963) points out that these animals prefer leaves of trees and shrubs to grass. Meadows would also show up in pollen diagrams in a similar way to the heavily grazed forest type, only with an early rather than late loss of AP values.

Clearance for vegetable resources

Clearance for the production of edible vegetation may well have been more intensive, entailing the removal of trees to allow more light to the ground. This would allow the establishment of a range of herbs, most of which would have been edible. Edible herbs could include the Leguminosae (eg Lathyrus, the pea genus), the Chenopodiaceae (eg Chenopodium album, fat hen), the Compositae (eg Taraxacum, dandelion), Potentilla anserina (silverweed, once cultivated for its edible root (Mabey 1972)), Urtica dioica (stinging nettle) and the Cruciferae (eg Sinapis arvensis, charlock). The subsequent invasion of Corylus would then provide nuts, and it is possible that coppicing was practised to encourage nut production.

Other shrubs which quickly invade clearances, such as Prunus spinosa, would be encouraged for their fruits. Thus clearances would be maintained at different stages of re-colonisation to supply different types of vegetable resources.

In clearances of this nature, hazel pollen values would always increase, due firstly to better flowering and then to colonisation into the clearance. AP values could decline if the area cleared was large. Ruderal and weed pollen should be plentiful and varied, particularly in the early phases, if resolution is good enough to allow phasing to be seen.

Clearance for other purposes

Evidence of man's activities may not be confined to discrete clearance phases, but any management technique which delays regeneration of woody vegetation. Indeed Vuorela (1973), in studying the pollen rain around cultivated fields, found that transportation and habitation were better represented in the pollen rain than the cultivation itself.

Unless a camp site was occupied over several seasons, any signs of habitation in pollen spectra would be ephemeral. Camps would be situated either in the shelter below open woodland or hazel scrub, and probably would not entail much clearance. However, in an area as densely populated as the central Pennines, there would be some indication of settlement, usually in the form of nitrophilous plants such as Urtica dioica and Sambucus nigra which invade the disturbed ground around human habitations. Fire-cleared tracks, or paths made by driving of herds, may show up in diagrams by the constant appearance of ruderals such as Chenopodiaceae, Artemisia, and Plantago major/media type (Behre 1981). However, it is impossible as yet to distinguish fire-cleared tracks from a) naturally trampled pathways, or, b) the presence of 'natural' refugia for such species in the open woodland of the uplands.

Glading is the harvesting of natural hay crops from under open woodland (Edlin 1970). If practised each year extensively it can have a great effect on the density of the woodland, as tree seedlings are usually taken, and thus regeneration is curtailed. Unfortunately, it is extremely difficult to document glading from pollen diagrams, a gradual loss of AP values over time being the major result of such an activity.

Thus, activities other than clearance for browse or vegetable resources are extremely difficult to document. However, where pollen diagrams show clearance phases, the existence of both nearby habitation and well-used trackways may be assumed. Glading, on the other hand, may be unlikely to occur given an annual round which demands movement to low ground in autumn. However, hay may be gathered for use late in the season. Evidence for such a practise was found at Oakhanger, Hampshire, where ivy was laid out either to attract game or to feed their herds (Rankine and Dimbleby 1960). Generally the detection of these types of activity waits upon the evolution of more sophisticated models for sampling and interpreting the pollen data.

To summarise the discussion so far:

1. The advantages of burning are manifold, not least in that it is an effective management technique producing good results from little effort.

2. Evidence for clearance in pollen diagrams demands specially designed sampling techniques and knowledge of the manifestations of different types of clearance.

3. 'Clearance indicators' have to be used with discretion, and are only of use when they form an ecologically <u>and</u> economically viable pattern.

4. Some management techniques are studied in terms of their representation in pollen diagrams.

Evidence for Clearances in the Soyland Moor Diagrams

Bearing in mind the possible misinterpretation of clearance indicators as discussed in the last section, the evidence presented below has to remain within the realms of possibility rather than fact. This evidence may possibly refer to natural rather than anthropogenic clearance, and, where possible, alternative explanations are discussed. The anthropogenic interpretations are based firstly on the density of the Mesolithic occupation of the area: if Mesolithic clearance is indeed a reality, then the central Pennines is the most likely area in which to find the evidence. Secondly, the evidence had to be interpretable as a sound economic strategy given our present knowledge of the Mesolithic, before an anthropogenic explanation was postulated. Thus, while this evidence remains speculative, the anthropogenic interpretation may stand unless other equally sound interpretations are found for these data. It is perhaps pertinent to point out at this stage that much of the following evidence supports the hypothesis that Mesolithic man manipulated his environment. However, it does cast doubts on the hypothesis that fire was the major mechanism employed. Thus, the terms 'clearance' and 'burning' are not interchangeable as they are within the original hypothesis (Simmons 1969, Mellars 1976b). The evidence for clearance is discussed in chronological order. Some evidence appears in only one diagram, other evidence can be seen across the whole of the mire. The discrete appearances in one diagram only underline the problem of attaining a full and accurate picture of clearance activity from pollen deposits (see also Edwards 1979a), and it is probable that the evidence given below under-represents the activity of Mesolithic man in the surrounding area.

The first Potentilla peak

SMC 623-592 cm (c 5950-5200 bc)
SMD 598-563 cm

This phase was termed the '*Potentilla* peak' after its most noticeable feature: a long period of high *Potentilla* type values. The duration of this peak is taken as the parameters for the clearance phase, which occurs during zones 3 and 4 of both SMC, and SMD. The clearance at both sites is contemporary.

Description

Unless otherwise specified, the description applies to both SMC and SMD.

Trees: This clearance phase occurs early, during the pine maximum, and arboreal values are rising until the end of the phase, when a sharp but short decrease is seen. Only the Ulmus values decrease slightly for the duration of the clearance.

Shrubs: Corylus values show the largest decrease of any pollen type. They decline gradually through the Potentilla peak, starting at 50-55% TP, and ending the peak at 20-25% TP. Prunus occurs at both sites during this phase, and Ligustrum appears at SMD.

Dwarf Shrubs: Calluna values are relatively high in the preceding period, but this phase marks a large increase of its values (up to25% TP). These high values are sustained until the end of the phase.

Herbs: At SMC, the herbaceous pollen values rise through this phase, and peak at the end (from 4% to 35% TP). Gramineae values follow this pattern, starting at 2% TP and ending at 20% TP. The highest values occur when Potentilla type values are declining. At SMC, the Potentilla type values reach up to 7% TP and Succisa pollen is continuously present. The curve for Succisa starts prior to that of Potentilla type, and ends in the middle of the Potentilla peak. Melampyrum pollen shows continuous presence just before Potentilla rises. SMD follows a similar pattern, but all these three herb types, and possibly also Gramineae have a bimodal distribution. This is most likely to be an effect of any slight mismatching when joining the two series of cores that comprise SMD. Compositae types become abundant at both sites. Valerianella occurs once at both sites. This type is rarely found in pollen diagrams (Godwin (1975a) mentions two records only); it may possibly have been confused with the Anthemis type of Compositae which is abundant during this phase, and its 'identification' at the same period in two separately counted diagrams mere coincidence. Many of the other herb types are only found, or are common at, one of the two sites. This is either because of the poor pollen dispersal of most herbs, or a reflection of the differing herbs found at different areas around the bog. Chenopodiaceae, Cruciferae and Umbelliferae are common at SMC, while Rumex acetosa/acetosella is common at SMD. Lotus and Spergula type are confined to SMD. Pteridium values increase at both sites.

The clearance indicators

The decrease in Corylus values is unusual, but by no means unique, in clearance phases (it occurs, for example, at Aros Moss, south-west Scotland (Nichols 1967), and Newferry, Co. Antrim (Smith and Collins 1971)). This may indicate that hazel scrub rather than trees themselves were being cleared. The continued decrease in hazel values would then be a result of grazing pressure. Any subsequent loss of trees within the cleared area would be masked by the continued invasion of trees onto the plateau at this period. The appearance of Prunus and Ligustrum pollen is most likely to represent the woodland edge around the clearance, where the extra light allowed shrubs to flourish. The increase in *Calluna* values is a common phenomenon during and after clearance phases on soils with a tendency to acidity (eg Iping Common, Sussex (Keef *et al* 1965)), as removal of woodland or scrub cover allows *Calluna* to invade (Dimbleby 1961b).

The gradual increase in herb values suggests that the cleared area increased in extent during the phase. However, this could also be explained by the greater pollen production of grasses, which increased only towards the end of the phase. The continuing decline in hazel, and the sudden decrease in AP values at the end, may indicate some increase in area or may suggest that seedling regeneration was stopped by intensive grazing. The appearance of the Chenopodiaceae, Umbelliferae and *Rumex acetosa/acetosella* may be unconnected to the clearance as they appear sporadically through the diagrams, but their slightly more common occurrence suggests that they took advantage of the clearance to expand. *Potentilla* type, *Succisa* and *Melampyrum* are the three major herb types which characterise this phase. They are all common to many clearance phases (Mamakowa 1968). In the case of *Potentilla* type there is a problem in distinguishing which species was involved. The species most likely to be associated with high Gramineae values is *Potentilla erecta.* This species is occasionally found on mires, but is more common in grassland on light, acidic soils (Clapham, Tutin and Warburg 1962). *Potentilla anserina* is also a possibility, as it is common on pastures which are heavily trampled. Turner and Hodgson (1983) found *Potentilla* type occurred most often with clearance phases associated with high Gramineae values in the northern Pennines. In the North York Moors, it is also common to many clearances, for example, Glaisdale Moor (Simmons and Cundill 1974a) and Tranmire Slack (Jones 1978). *Succisa pratensis* is found today mainly in wet pastures (Behre 1981). Vuorela (1973) states that it is less likely to refer to clearances because its heavy pollen travels only short distances, but Mamakowa (1968) noticed increased *Succisa* occurrence in deposits after deforestation. A number of clearances elsewhere underline its association with *Potentilla* type (see, for example, the North York Morres sites mentioned above). *Melampyrum* (which usually refers to *M. pratense* (Godwin 1975a)), occurs most often today in woodland (Clapham, Tutin and Warburg 1962). However, many researchers note its proclivity for disturbed habitats, especially those which have been burnt (eg Danielsen 1969, Mamakowa 1968, Iversen 1949). In the North York Moors, *Melampyrum* occurs at practically every clearance phase (see, for example, the papers by I G Simmons and R L Jones). Behre (1981) states that it occurs, along with *Pteridium,* in grazed areas. *Pteridium* does increase at these levels, and both Oldfield (1963) and Simmons (1969), use this increase as an indicator of grazing. The occurrence of *Spergula* type at SMD is interesting, as it does not usually occur before the Elm Decline (Godwin 1975a). However, it also occurs at a clearance phase at Ewe Crag Slack Site in the North York Moors (Simmons *et al* 1975). Here it is also associated with *Melampyrum.* The type includes *Spergula arvensis,* a calcifuge weed of arable land, and *Spergularia rubra,* a calcifuge weed in open, sandy habitats (all other *Spergularia* species are confined to coasts, (Clapham, Tutin and Warburg 1962)).

Thus this clearance is characterised by loss of hazel scrub and ruderals mainly associated with pastures.

Interpretation and correlations

All the features of this phase suggest a long period of intensive forest grazing. Initial clearance of hazel scrub and undergrowth may have been carried out by fire. (*Melampyrum*, which is often associated with burning, occurs only at the very early stages of this phase.) If burning did take place it would serve to attract the herbivores initially. The evidence thereafter, along with the loss of *Melampyrum*, conforms with the model of intensive forest grazing discussed earlier in this chapter, with some form of confinement of the herds. Under this form of management, fire would no longer be necessary. Confinement may take the form of fencing: hazel hurdles between the surrounding trees, for example, would not be too time-consuming to erect. Use of herdsmen is possible, although less likely with 'undomesticated' animals (although precisely which animal was being herded, and to what extent they had been 'domesticated' is unknown). The evidence which demands the use of some form of confinement rests on the development of a grass sward, which most herbivores find less palatable than shrub or leaf browse (Heybroek 1963). Given a choice, herbivores would move out of such an area. This hypothesis of 'intensive management' is corroborated by the lack of hazel regeneration, as it will only decline under heavy grazing pressure (Dimbleby 1978). If periodic burning was taking place to maintain small areas with maximum browse quantities to attract herbivores, then not only would hazel be able to regenerate in these comparatively lightly grazed areas, but the invasion of hazel into abandoned clearings would maintain their values in the pollen diagrams. The smooth profiles of all the clearance indicators also suggest one long phase of local clearance, rather than a series of small ones. The abnormally high values of the ruderals, particularly *Potentilla* type, suggest that the smooth curves are not the product of distance. At SMC, the peak of Gramineae values occurs after the peak of *Potentilla* type, suggesting that a grass sward was a later development, and possibly covering a larger area than the original clearance, although the evidence for this, as discussed previously, is equivocal. The high values of the pasture-type weeds in the earlier stages of the clearance may be indicative of over-grazing and trampling, so that a larger area was cleared to relieve the pressure. At SMD, Gramineae values reach their highest values during the *Potentilla* peak itself. This may be a factor of mismatching of the cores, but may also be a true reflection of heavier grazing pressure than at SMC, allowing an earlier development of grass sward.

Evidence from the North York Moors shows that clearances for grazing are common in Zone VI. Some sites show the same sequence of high weed values before high grass values as at SMC (eg Ewe Crag Slack sites A and B (Jones 1978) and West House Moss (Jones 1976b)). Other sites show that *Potentilla* and/or *Melampyrum* values peak with the grasses (eg Glaisdale Moor (Simmons and Cundill 1974a)). Whether these latter sites mirror the true state of development within the clearances, or whether the evidence is compressed due to slow peat growth is impossible to determine without C14 dates. Certainly, most of these North York Moor clearances appear to be much shorter than those of the central Pennines. The greater values of *Melampyrum* when compared to *Potentilla*, and the increase in *Corylus* values tends to suggest less intensive management, with small fired clearings used as a basis.

The increasing AP values during this phase at Soyland are explained either by the masking out of local clearance by the increasing AP values generally, or may point to clearance of shrubs only. The retaining of the standards could be a mere side-effect of the low intensity burning used initially, or could reflect a deliberate policy. The open nature of the woodland at this period would mean that the retention of the mature trees would not greatly impair visibility, and keeping the trees would stop the rapid leaching which occurs with deforestation (Fleming 1972). The increasing values of Calluna underline the acidity of the soils, and it is likely that the Mesolithic groups were not wholly unaware of the effects of drastic deforestation on the subsequent vegetation growth (Bradley 1978). The sharp loss of AP values towards the end of the phase could merely represent the threshold of grazing pressure which prevents tree regeneration, but could also be a reflection of incipient podsolisation under the Calluna heath which was progressively invading the clearing. This, along with the removal of nutrients by the herds, could effectively prevent seedling establishment (Dimbleby 1952b). It is just after the initial loss of AP values that the clearing was abandoned, which may corroborate the latter hypothesis, as a loss of trees within the clearing alone would not necessarily greatly reduce its carrying capacity. With increasingly acid soils, grazing would be of poorer quality, and the effort of maintaining the clearance too great for the results.

The slight decline in Ulmus values for the duration of this phase is more difficult to explain. One hypothesis (put forward as a possible explanation for the 'real' Elm Decline) suggests that more fertile areas would be picked out for clearance by the greater quantities of elm growing there (eg Nichols 1967). If this hypothesis is correct, then the elm trees would probably have been killed by ring-barking or felling. Burning is less likely, as high intensity or repeated firing would be needed to kill mature trees and is unlikely to be so selective. The lack of charcoal (verified by J H Tallis) in any great quantity in the Soyland Moor cores corroborates this. Ring-barking may be indicated by the almost constant occurrence of Hedera during this phase at SMC: Garbett (1981) points to the prevalence of flowering ivy on dead elm standards today. However, the data does not fully bear out this hypothesis. It would demand that elm was growing in almost pure stands, otherwise the values of other tree species would also diminish (there would be little point in only killing elm within mixed woodland). Also, the sharp loss of AP values towards the end of the clearance (mainly of oak and possibly alder) would then be difficult to explain. Perhaps a more viable hypothesis is the lopping of elm for fodder, during late summer for use in autumn (eg Troels-Smith 1960), or even the selective browsing of the herbivores, as elm produces the most palatable fodder for herbivores (Troels-Smith 1960). Even if other tree species were lopped or browsed, elm takes seven to eight years before the broken branches produce flowers again, compared to two or three years for most other species (Heybroek 1963). Thus, the evidence of increasing AP values and loss of elm values can be interpreted either as a loss of trees in the clearance or as forest grazing.

The extent of the area cleared is more difficult to ascertain. The fact that he phase is clearly seen at both SMC and SMD could mean either that it was not a local event, or that one large area as cleared, or more than one clearance was maintained. The high values of insect-pollinated herbs and the continuous records of heavy herb pollen (eg *Succisa*) which does not travel far, strongly suggests that clearance was local, taking place in the immediate vicinity of the mire. Evidence from Robinson's Moss (Tallis unpublished) may help to decide which of the remaining hypotheses is likely to be correct. Although fully 20 km from Soyland, Robinson's Moss shows an extraordinarily similar phase which has a date of 5725 ± 65 bc for the peak of *Potentilla* type values. This makes it contemporary with the phase at Soyland. The values for *Rumex acetosa/acetosella* are higher at Robinson's Moss, and *Epilobium* occurs instead of *Melampyrum*, but apart from this, the phase is identical. This evidence appears to suggest that large tracts of the central Pennines were used for an intensive form of herding, very similar to the form of pastoralism practised in the Bronze Age (Fleming 1972). If this was the case, then the use of fencing would not have been necessary. If large tracts of the area were under open woodland with a grass sward undergrowth, then herbivores would not be able to find more attractive browse. Herdsmen would stop them from moving out of the area altogether. More diagrams covering this period from the central Pennines are needed before any firm conclusions can be made.

As discussed, the evidence from other areas, particularly the North York Moors, shows that clearance for grazing purposes was common in Zone VI. Jones (1976b) suggests that herding on a large scale was practised over much of the North York Moors at this period. However, when compared to the central Pennines, one essential difference can be seen, in that AP values decline while *Corylus* values increase at most of these North York Moor sites (eg Ewe Crag Slack Site C (Simmons *at al* 1975) and Tramire Slack (Jones 1978)). This suggests, on the one hand, that more drastic measures were used, with the removal of the tree canopy to a greater extent than may be the case at Soyland and Robinson's Moss. On the other hand, the increase in *Corylus* values may indicate that the management technique itself was confined to successive burning, a much less intensive technique than that postulated for the central Pennines.

The abandonment of the clearance

The abandonment of this form of management at c.5300 bc is marked by the rapid decline in herbaceous pollen values, and the recovery of the AP values. However, the peak in *Calluna* and other ericaceous species for a period of c.200 years after the abandonment suggests that trees and shrubs did not rapidly invade all the cleared area. A short peak in the *Betula* values is seen during and after that of *Calluna*. *Betula* is the tree most likely to be able to invade acid heathland, and the subsequent loss of *Calluna* could then be a result of shading (Gimingham 1972). Removal of *Calluna* would allow the soils to recover and possibly to re-form a mull humus profile under *Betula* (Miles 1978). Only then could *Corylus* re-invade the cleared areas, as is seen by the slow recovery of its values. At SMC, this recovery of hazel occurs with the decline of pine values, and thus conforms with the pattern seen at many British sites (Godwin 1975a). However, at SMD, hazel values actually begin to rise during the pine maximum, suggesting these two events are in fact unconnected at Soyland. It is, therefore, likely that this secondary *Corylus* rise is, as Smith (1970) suggested, a result of man's activities rather than natural vegetation development. However, this is applicable so far only to the central Pennines, where *Corylus* marks the recovery from, not the beginning of, a clearance phase. This sequence of events seems to uphold the model of an inverse relationship between *Calluna* and *Corylus* (rather than *Sphagnum*) that was discussed in the previous chapter.

Alternative Interpretations

The evidence of this period at both Soyland Moor and Robinson's Moss (Tallis unpublished) points to a phase of grassland. However, it may be possible to interpret this merely as an expansion of rough wet grassland around the edges of the mire at the expense of the hazel scrub. This could take place at a point when the mire itself was about to expand, and ground conditions in the immediate vicinity became wetter and could no longer support trees or scrub. The increase in *Calluna* values during, and especially afterwards, would then represent the extending drier margins of the mire itself. The presence of ruderals would be explained by the use of this open area by herbivores (and of course, man would then have taken advantage of the situation).

There are no convincing reasons for rejecting this hypothesis outright. However, there are some points which are difficult to explain:

1. There is no evidence in the pollen and spore values, particularly Cyperaceae and *Sphagnum* and wetland herbs, that the bog was extending at this period.

2. If this phase represented a local hydrological change, it is curious that it also occurs at Robinson's Moss, 30 km away.

3. The decrease of hazel and elm values in association with a general increase in AP values is difficult to explain, unless the mire was encroaching upon a pure hazel and elm woodland occupying flushed areas in the immediate vicinity. The sharp decrease of AP values at the end of the phase may then be attributable to the extension of the mire beyond the flushed areas into the surrounding woodland. However, if this is the case, it is difficult to find an explanation for the subsequent recovery of the hazel and elm values, as no flushed areas would then be available for their re-establishment.

While this explanation remains viable, the anthropogenic explanation appears to be more complete, particularly in the period of regeneration. Further research at different mires in the area (if they exist) would be necessary to help to decide the explanation.

Perspective

If this phase was anthropogenic in origin, then it occurs relatively early in the postglacial at c.5950 bc. The evidence is indicative of a more intensive form of management than has been expected of Mesolithic peoples. Either extensive areas were kept open under forest grazing, or many fenced areas were used to control the herds. The clearance lasts for over 600 years, being abandoned when the effects on the soil were just short of catastrophic. Indeed, in the pollen source area of SMD, *Calluna* values remain high enough to indicate permanent areas of *Calluna* heath, under which podsolisation would have rapidly occurred, leaving the areas much less productive than previously. However, in the pollen source areas of SMC and Robinson's Moss, the very low values of *Calluna* indicate that soils here recover completely. Certainly, the general recovery of *Corylus* and AP values indicates that only pockets remained too acidic for any species other than those of the Ericaceae.

If this type of management was occurring over extensive areas of the central Pennines, then it may well have had a profound effect on the establishment and spread of the invading species during this dynamic period of the postglacial. The possible effects on *Corylus* have been discussed. Indeed, the very open nature of the woodland into which *Corylus* regenerated must have been due in part to the intensive grazing. The role of pine in the area may also have been affected by this management. Rackham (1976) suggests that the loss of pine at the BAT was due to clearance by Mesolithic man as pine is the easiest tree to kill by burning (Edlin 1970, Carlisle and Brown 1968). However, as with *Corylus*, the effect of Mesolithic man here is directly opposed to the prevailing ideology: pine remained in the central Pennines for a longer period than elsewhere, despite the intense activity of man. Smith (1982) does indeed suggest that pine can be advantageously affected by burning as it regenerates well in burnt areas. However, at Soyland and Robinson's Moss, it seems unlikely that fire played an important role in the clearance, as it was apparently maintained solely by grazing pressure. Also, this very grazing pressure would have checked the establishment of the pine seedlings (witness the deer fencing today around Scottish pine plantations). The explanation of the extended pine maximum may lie in the isolation of the pine stands on the well-drained rises of the central Pennine plateau, as postulated in previous chapters. If clearances were fenced, then these areas were not used. If, however, much of the plateau was under intensive forest grazing without fencing, then the explanation would lie in the less attractive browse available under dense pine stands. This is due to the heavy shade of dense pine stands, the occurrence of mor humus under pine, and the greater leaching of the well-drained rises as compared to the rest of the plateau. The slow spread of oak could be directly explained by grazing pressure, and this could indirectly have prolonged the dominant position of pine, as oak would then be less able to expand into the pine stands. Therefore, clearance could have prolonged the dominance of pine negatively, through the lack of grazing, and positively, in the grazing of pine's competitors.

Thus, this clearance phase could have had profound effects on the vegetation succession in the area. It was an intensive form of management, comparable to that of the Bronze Age, and possibly even comparable with the parklands which constituted the medieval hunting forests. The dates for this phase coincide with the early part of the Late Mesolithic (Mellars 1974). This phase may then represent the economic pressure which accompanies culture change (eg Jarman 1972).

The Second Potentilla Peak

SMC 560-505 cm (c 4650-3015 bc)

? SMD 512-513 cm

Description

SMC: The expression of this phase is more complicated than the first Potentilla peak in that a wider variety of herb pollen is recorded, AP as well as Corylus values decline and two major phases of clearance occur, but the two sequences are basically similar. Table 2 outlines the pattern of the clearance indicators.

SMD: The clearance phase of SMC may just occur in a vestigial form at SMD. Herbaceous values are generally very low (under 5% TP), and the Gramineae values do not rise at all during the period. However, coincident with Phase 1 at SMC, two contiguous levels at SMD (512-513 cm) show Potentilla type values of 1.5% TP. Corylus values decline at these two levels only and Ranunculus acris type and Jasione montana occur here (the latter occurs sporadically through this zone). Betula values increase slightly afterwards, as at SMC. The Calluna profile too shows a broadly similar pattern to SMC, although in a much less erratic fashion. Two features possibly indicative of clearance are seen at SMD which do not occur at SMC. Firstly, pine values, which were just beginning to decline, drop sharply just before the appearance of Potentilla at 512 cm, and do not recover. Secondly, Pteridium values are higher and much more erratic at SMD, whereas they remain low for much of the episode at SMC. At SMD, the picture is confused by the occurrence of a short local clearance of an entirely different character on the SMD 4/5 boundary at SMD 500. This coincides with Phase 2 of the SMC clearance.

Interpretation and Correlations

The interpretation of this as a clearance rests on the arguments presented for the first Potentilla peak. However, its occurrence in the diagrams is largely confined to SMC, suggesting it was a much more local phenomenon, and, may therefore indicate a small fenced-off clearance (in the first grazing phase, the sheer extent of the clearance may have kept the herds in the general area. If a small area only was maintained as grassland, then the herbivores would not remain there unless some form of confinement was used) . The phase is divided into two major episodes. Again, the major increase in grass values occurs after the Potentilla peak, but this time AP values decline and remain low for the duration of the clearance (including Ulmus) along with the decrease in Corylus. This decrease in AP values may be simply explained by the greater extent of the woodland at this time. More trees would be interspersed through the Corylus scrub, and the initial clearance, either by burning of confined grazing, may have resulted in the loss of both trees and scrub. However, the initial clearance must then have been of a high intensity to have killed mature trees.

Phase 1 of the clearance seems to have been abandoned for a period of 200 years at the end of the fifth millenium bc. This is seen firstly in the increased values of Calluna, then by Betula and Corylus, along with the decline of the pastoral indicators. It would appear that temporary soil exhaustion (seen in the heathland

Pollen type values	Phase 1 Clearance: 560-542 cm (SMC 4) c 4650-4250 bc	Phase 2 Regeneration/ acid heath: 536-535 cm (SMC 4/5 boundary) 4160 ± 40 bc	Phase 3 Clearance: 536-527 cm (SMC 5) c 4040-3800 bc	Phase 4 Rough grazing 527-505 cm (SMC 5) c 3800-3015 bc	Phase 5 Abandonment 505 cm (SMC 5) c 3015 bc
Arboreal	Declines	Rising, peaks at end	Declines	Gradually rising	Peaks
Betula	Declines	Increases	Declines	Increases	Slight decline
Fraxinus	Sporadic	Continuous present	Increasing	Peak at end of zone	Declines
Corylus	Declines	Rising, peaks at end	Declines	Declines	Peaks
Crataegus	-	-	-	One grain towards end	-
Calluna	Rising from very low values	Peak (20% TP)	Erratically declining	Erratic	Declines
Other Ericales	Small increase	Decline	Very small peak	Continuous presence	Sporadic
Empetrum	small	peak	-	-	-
Herbaceous	High (end of long period of high values)	Declines	Increasing	High	Rapid decline
Gramineae	High, but slightly declining	Small peak	Increasing	High	Rapid decline
Potentilla	Peak (90% TP)	Decreases to continuous presence	Peak (4% TP)	Decreases to continuous presence	Sporadic
Melampyrum	Present at middle of *Potentilla* peak	Sporadic	-	-	-
Succisa	Continuous presence	Sporadic	Sporadic	-	-
Rumex acetosa	Small peak	Continuous presence	-	-	-
Chenopodiaceae	Continuous presence	Continuous presence	Declining	-	-
Plantago lanceolata	One grain	-	-	-	-
Pteridium	Low	Slight increase	Decline	Rising, with maximum of 14% TP at end	Declines
		Decline	Rising, with maximum	Declines of 14% TP at end (505 cm)	

TABLE 2 Description of the Second Potentilla Peak at SMC 4/5

development) caused Mesolithic man to move the herds away until this process was reversed by the establishment of trees and shrubs, under which the soils could regenerate (Dimbleby 1952b). Clearance was resumed at c. 4040 bc, at the point when deciduous woodland was becoming more important than pine. The maximum Gramineae values do not occur until c. 3800 bc, and coincide with high Pteridium values, possibly indicating that Phase 4 was a period of rough grazing (Oldfield 1963). In the early phases, Pteridium was more representative of the regeneration phase, possibly occurring within the Calluna heathland.

Despite the fact that this clearance lasts for over 1500 years, much longer than the first one, it was a much more localised event in the pollen source area of SMC. This is seen in the much more erratic profiles of the clearance indicators, and its vestigial occurrence only at SMD. This latter is difficult to explain. No major clearance indicators are present, and the broadly similar, but less sharp, profiles of Betula, Corylus and Calluna tend to suggest that the SMD diagram is merely reflecting a clearance much closer to SMC. However, the sudden loss of Pinus values at SMD is difficult to fit into this hypothesis, unless the Pinus profile changes are unconnected. Clearance at the SMD pollen source area within the pine stands themselves is a possible, but unlikely, explanation.

Whatever the explanation of the events near SMD, the main thrust of the clearance was centred in the pollen source area of SMC. Evidence from Robinson's Moss (Tallis unpublished) may, however, indicate that clearance was occurring at a number of areas in the central Pennines. At Robinson's Moss the evidence for clearance is similar to the vestigial occurrence at SMD, with a slight increase in Potentilla type and Rumex at two levels only, and a drop in Pinus values just before the herbs occur. Afterwards, instead of an increase in Betula values, Calluna values increase sharply for one level only. This is contemporary with Phase 3 of the SMC clearance, rather than Phase 1 as at SMD. On its own, the evidence at Robinson's Moss could not be decisively interpreted as clearance activity, but in conjunction with the Soyland evidence, may now be interpreted as some reflection of nearby clearance activity. Taking the evidence from the three sites, it is possible that the extent of the clearance was not smaller than during the first Potentilla peak, but this time was merely centred away from the mire edges, so that its representation in the mire may not be clear.

The Abandonment of the Clearance

Towards the end of Phase 4, relaxation of the grazing pressure may be seen in the gradually increasing AP values, particularly Betula and Fraxinus. Crataegus also occurs. This indicates the invasion of parts of the cleared areas by those trees and shrubs quick to respond to the increased light, and previously kept out by grazing pressure. The high values of Pteridium in Phase 4 may indicate that the grazing had become degraded. The erratic Calluna values point to areas that were reverting to heathland. Thus the increasingly poor returns from the expenditure of effort in maintaining the cleared area may explain its abandonment. It is possible that areas further from the mire were then subjected to clearance, too far away to be represented in these diagrams.

NAP values rapidly decline when the area was abandoned, both Corylus and AP values rising as the area reverted to woodland. The slight decline in both Betula and Fraxinus values signals the full regeneration of the cleared areas.

Perspective

The second Potentilla peak represents intensive clearance at a localised centre for a period of two millennia. It spans the period when pine ceased to be the dominant tree and oak and alder woodland extended rapidly. The first indications of this episode appear at the end of a period of high herbaceous values (the lowest herbaceous values seen since the end of the first Potentilla peak actually occur during the temporary abandonment of the clearance (Phase 2)). Thus, this episode would appear to occur at a time when there is evidence for a large area of open ground around the mire, which should indicate clearance was in fact not necessary. However, Cyperaceae pollen contributes to much of these high herbaceous values in the period prior to clearance, and may be representative of the mire surface vegetation, rather than open ground on the plateau top. Besides, clearance activity may not have been triggered by the absolute amount of open ground but by the woodland type. While AP values themselves do not rise during this period, pine was rapidly disappearing from the area, and oak and alder were beginning to dominate. If pine was confined to areas not subjected to clearance, then the increasing density of oak and alder woodland all over the plateau would have reduced the area of open ground utilised by the herbivores, and thus clearance would be necessary.

If this latter hypothesis is correct, then the second Potentilla peak is comparable to a large number of clearances which are clustered around the BAT (eg Tarn Moss, Malham (Pigott and Pigott 1963), Stump Cross, Grassington (Walker 1956) and Glaisdale Moor and Bonfield Gill Head in the North York Moors (Simmons and Cundill 1974a, Simmons and Innes 1981)). These clearances all occur as deciduous forest expands. However, the dating of this clearance at SMC is contemporary with clearances which take place in the middle of the Atlantic period. These are also numerous and include Tranmire Slack in the North York Moors (Jones 1978), Valley Bog in the north Pennines (Chambers 1978) and Borth Bog in Cardiganshire (Moore 1968). These clearances may have been the result of renewed population pressure, forcing more intensive economic measures. Thus the second Potentilla peak may represent drastic measures undertaken under the dual pressures of increasing population, and the degeneration (in strictly procurement terms) of the environment as deciduous forest invaded.

Again, it is perhaps possible to interpret this phase as a natural clearance, this time only along one edge of the mire. However, the vestigial representation at SMD, and particularly at Roginson's Moss, is difficult to explain in these terms, and the more complex vegetation development is also more readily explainable in anthropogenic terms than by an extension of the mire.

Clearance at SMD 500

3870 ± 95 bc

This is a very circumscribed clearance, visible at one level only and not seen at all at SMC. The changes in the pollen values are very sharp, and the phase is clearly picked out in the summary diagram.

Description

Trees: 1. AP values decline from 40-34% TP at this level.

2. All tree values are affected except Alnus and Fraxinus.

3. Fraxinus pollen has appeared sporadically in earlier zones. 500 cm marks its first continuous presence until 495 cm where it disappears again.

4. Ulmus values decline at 504 cm and remain low until 500 cm after which they fully recover.

5. One grain of Fagus appears at this level.

Shrubs: 1. Corylus values increase at 499 cm from 40-55% TP, drop to 35% TP at 500 cm, and then revert to 40% TP.

2. One grain of Prunus occurs at this level.

Herbs: 1. Gramineae values rise from 2-15% TP at this level.

2. Three cereal grains occur at this level (see Appendix III for identification).

3. Plantago lanceolata, occurring only sporadically before and after 500 cm, has values here of 3% TP.

4. Rumex acetosa/acetosella values reach 1.5% TP.

5. Rumex obtusifolius type, Artemisia, Plantago major, Umbelliferae and Compositae all occur at 500 cm.

6. Pteridium values are generally low but erratic around this level. At 500 cm, they increase to 4% TP.

Aquatic and wetland species: 1. Cyperaceae values increase from c. 2-5% TP.

2. Sphagnum values rise from 500 cm, reaching 110% by 496 cm before declining.

3. Caltha, Potamogeton and Ranunculus cf trichophyllus occur, as well as the more common Filipendula and Lysimachia vulgaris type.

SMD 500 (repeat): Because this unusual clearance occurred at the edge of a core, a further core was extracted 2 m from SMD at 525-475 cm below the surface, to ensure that this phenomenon was not the product of contamination. At exactly 500 cm in the repeat core, a very similar profile appeared, showing almost the same clearance indicators. One cereal grain was found. The major difference was one of scale: Corylus and AP values did not decline as much, and Gramineae values do not rise above 5% TP. Appendix IV gives the full figures for SMD (Repeat) 503-497 cm.

Clearance Indicators

Cerealia: Cereals are, of course, the best indicator for man's activities. The problems here lie in the existence of some other grass species (eg *Glyceria*) with similarly large grains; the inability to identify the cereal genus due to the poor preservation of the sculpturing and annuli, and the difficulty in accepting such an early appearance of cereals. However, the appearance of a total of four grains from a doubly sampled level in association with so many other clearance indicators and in such a sharply defined manner justifies the acceptance of these grains as cereals.

Fagus: Because *Fagus* does not normally occur until after the Elm Decline (Simmons 1969), and because it occurred at SMC 1 as contamination, this one grain, even more than the cereals, seemed to cast doubt over the validity of the clearance level. However, the double sampling corroborated the existence of the clearance, and the levels above SMD 500 cm show pollen profiles quite consistent with the regeneration of the clearance. Also, the date is acceptable for the pollen profiles, and is confirmed by the date of 4160 ± 40 bc for the same period at SMC. This date for the first appearance for *Fagus* is certainly early, but not unique in British pollen diagrams (Godwin 1975a). Also, *Fagus* pollen often first appears in diagrams with evidence for clearance, as it responds rapidly to the removal of competition (Simmons 1981), and its erratic flowering is helped by the extra light (Godwin 1975a). This implies it was present in British woodlands some time before it is represented in the pollen diagrams. Accordingly, the record is accepted as valid, although early and on soils prone to acidity.

Plantago species: To a large extent *Plantago lanceolata* is associated with pasture (eg Vuorela 1970), yet it does not occur at Soyland during the long clearance phases for grazing, and here it occurs with cereal grains. This may be explained by its other associations with fallow land (Behre 1981), suggesting it invaded either after the clearance was abandoned, or during fallow phases if SMD 500 represents more than one season of cereal growing. *Plantago major* is more indicative of trackways than arable cultivation (Behre 1981, Sagar and Harper 1964), which would be compatible with an intensively used area.

Rumex species: Again these species tend to be more representative of pasture rather than arable land (Vuorela 1970, Behre 1981), but they occur also in disturbed ground (Behre 1981).

Generally, apart from *Cerealia* pollen itself, the clearance indicators are more usually common in pasture. However, the herbs associated with pasture in the Soyland diagrams (*Potentilla, Melampyrum, Succisa* and Chenopodiaceae) do not occur at all at SMD 500. Also, most of the so-called 'pastoral' indicators may be merely indicative of open ground. There are few obligate species for pasture (Behre 1981), whereas cereal grains are a very definite indication of arable cultivation. Pastoral and arable indicators found from modern pollen samples (eg Vuorela 1973) may not be wholly applicable to these early experiments with farming, as the economic system may not have been quite so neatly divided as today. There is no evidence for ploughing and the ground may have only been roughly broken up, leaving many of the ruderals which grow mainly in grassland.

Interpretation and Correlations

There are two immediate difficulties in accepting the presence of cereal grains at SMD 500. Firstly, the early date, at 3870 ± 95 bc. Cereals do not usually occur until the post-Elm Decline period, and SMD 500 cm occurs at the end of the pine decline. Secondly, the sharp pollen profiles and the lack of evidence for this type of clearance at SMC at this time, indicates that this clearance occurred very close to the mire and not in the valleys. That cereal growing was attempted at 389 m OD seems at first to be highly unlikely.

As to the first difficulty, that cereals do not usually occur until after the Elm Decline, there is an increasing number of sites which tentatively record cereals prior to this. The difficulty of accepting these early records is underlined by the many restrictions researchers place on these cereal records: identification is doubted, contamination suspected (as above); Lynch (1981) neatly side-steps the issue altogether by referring to her levels with cereal grains as 'Early Neolithic'. The sites which record early cereals are mainly Irish, where the rather different botanical and archaeological post-glacial history may make comparisons with Soyland tenuous. However, three English sites may record early cereals. The full list of sites with early records of cereals are:

1. Newferry, Co Antrim (Smith and Collins 1971):
   During a densely forested period of Zone VIIa clearance occurred, represented by loss of AP values, except Betula and Fraxinus, and Corylus. One cereal grain only.

2. Ballynagilly, Co Tyrone (Pilcher and Smith 1979):
   One cereal grain only, accompanied by rather vague indications of clearance, dated to c 3800 bc.

3. Dolan, Connemara (Teunissen and Teunissen-Van Oorschot 1980):
   One grain of Triticum occurs with high values of Plantago lanceolata in the immediate pre-Elm Decline period. AP values decline briefly, the brunt of the losses being borne by Ulmus.

4. Cashelkeelty Site I, Co Kerry (Lynch 1981):
   Two clearances with cereal grains are found: At 3895 ±100 bc, one grain of Triticum is associated with decreased AP values. This is followed by a long phase of clearance for grazing (until c 3550 bc). After a period of regeneration, two cereal grains (one of Triticum, the other Hordeum) are found at c 3400 bc, again associated with lower AP values, followed by grazing.

5. Rinsmoor, South Downs (Waton 1982):
   Little information is given about this important site, (summary diagrams only) where peat growth has been estimated at 1 cm per 3.84 years, which should give very detailed information about the progression of any clearance. The cereal grains appear at c 3550 bc, 350 years prior to the Elm Decline.

6. Tranmire Slack, North York Moors (Jones 1978):
   The first evidence for cereals here occurs just before

the Elm Decline. They are associated with high values of Plantago lanceolata, Rosaceae, Potentilla type, Succisa and Gramineae.

7. Oakhanger, Hampshire (Rankine and Dimbleby 1960):
This site is well-known for the huge quantities of Hedera helix pollen which has been interpreted as evidence for some form of herding in the Mesolithic. However, an earlier phase of this diagram shows a clearance level which is remarkably similar to SMD 500. In the article, Dimbleby dismisses the evidence as contamination, probably laid down through a root channel within the soil profile, a suggestion supported by signs of humic staining which may indicate the presence of such a root channel. However, the 'contamination' shows an ecologically sound representation of clearance, whereas contamination might be expected to produce a random, inexplicable spectrum. One cereal grain is accompanied by increased values of grasses, Pteridium, Rumex and Calluna. Both AP and Corylus decline, and curiously, one grain of Fagus also appears as at SMD 500. The major difficulty in accepting this as a real event is the date of pre-4350 bc, and here the Oakhanger data is presented merely for consideration.

Also, finds of carbonised barley and oat grains in a Mesolithic context at Unstone, Derbyshire, await radiocarbon dates (Courtney 1978). Thus, evidence for pre-Elm Decline cereals is growing, and despite the doubts often expressed by the researchers who find these grains, it would appear that their occurrence will soon be accepted as usual.

Most of the above sites are either in the lowlands or mid-altitude (eg Tranmire Slack at 195 m), as may be expected. Today, cereal growing in the Pennines is precarious above 230 m (Barnes 1982). However, this figure cannot be directly projected back into the past. The climate at c 4000 bc was certainly warmer (Lamb 1977), although it may have been wetter (Moore 1975a), and the existence of open woodland and scrub would have reduced the strong exposure factor which operates today in the largely treeless Pennines. Also, soils were more base-rich at this period, which is prior to the major expansion of Calluna. On a priori grounds, cereal growing would take place in the area where Mesolithic peoples spent their spring and summer months, which was most likely to have been the uplands. As a corollary to this, Mesolithic peoples would regard the uplands as the better environment compared to the lowlands, with a mental template of feast rather than the winter scarcity. If no annual round occurred, then the evidence must point to the uplands as being more densely occupied than the lowlands and the same argument applies.

The clearance at SMD 500 is most likely to represent just a few seasons of cereal growing. At SMC, one centimetre represents c 20 years according to the radiocarbon dates, but the faster growth of peat at SMD means that one centimetre represents a lesser duration, possibly nearer 15 years. This may mean that this clearance was an experiment with a new economic strategy that was beginning to be used in the lowlands of southern Britain about this time (Bradley 1978).

However, it is possible that other areas were being used in this way. The next clearance (at SMC 500), may indicate this, although no cereals were found. Cereal pollen does not travel far, and is often dependent on man's activities, such as threshing, to disseminate it. Vuorela (1973) suggests, from modern pollen experiments, that in fact transportation routes and habitation areas are more likely to produce pollen evidence for cereals. At SMD 500, all the evidence points to the cultivation area itself, but, of course, the cultivation could well have been placed near to habitation. Fencing or constant surveillance would be necessary to stop the crops being grazed: the situation of the cultivated area near to habitation would reduce this need, as herbivores would not frequent these areas. The appearance of Plantago major and Artemisia could support this hypothesis of cultivation around a settlement, as they are indicative of trampled areas. However, the lack of such nitrophilous species as Urtica dioica, which sporadically occurs through the diagrams, and Sambucus, which actually disappears at this point, may point to habitation further away.

When comparing SMD 500 with the other sites that have recorded early cereals, it can be seen as one of the clearest episodes of cereal growing that exists, apart from Cashelkeelty (Lynch 1981). At Cashelkeelty cereal production occurs at the start of two successive periods of clearance for grazing. At Soyland, clearance for grazing is registered at SMC over the period in which cereal growing occurred at SMD, implying that the economy is mixed. This could also be the case at Tranmire Slack (Jones 1978), where cereals occur with herbs that may be more representative of grazing (Potentilla type and Melampyrum). Here it is possible that slower peat growth is masking a sequence similar to Cashelkeelty, or possibly one large clearance with some areas for grazing, others for cereals, as some researchers postulate for the Neolithic 'landnam' clearances (eg Troels-Smith 1960). The Mesolithic clearances cannot be dual purpose, as areas too well-frequented by man would drive away the herbivores. However, the evidence from Soyland indicates more complex and intensive economic strategies may have been used: if the herbivores were fenced in, then clearances could well have been used for more than one purpose.

The Dolan site (Teunissen and Teunissen-Van Oorschot 1980) is interesting in that it is accompanied by a decline in Ulmus values. As such it conforms with clearances found at the Elm Decline proper. At Soyland, elm values fall at every clearance phase, but SMD 500 shows a slightly different pattern, with elm falling prior to the clearance, and is actually rising again when the cereal phase occurs. In this, SMD 500 can be compared to a number of sites at the VIIa/VIIb boundary, where elm declines prior to a 'landnam' phase (many Lake District sites show this pattern, (eg Oldfield 1963, Pennington 1975). No ready explanation for this phenomenon is forthcoming, either for the Elm Decline itself, or for this particular phase at SMD 500, where no other pollen profile changes with the initial decline in elm. The possibility of utilising elm leaves for direct human consumption (sometimes seen as a starvation measure (Dimbleby 1978, Mabey 1972)) before resorting to cereal growing is a possible, if rather eclectic, explanation.

The Abandonment of the Clearance

Unlike the other clearances at Soyland Moor, very few long term effects are registered, perhaps corroborating the experimental, short term nature of the episode. Fraxinus remains for a short period before both Corylus and AP values fully regenerate. The herbaceous pollen values quickly decline to their pre-SMD 500 low values.

The Calluna profile does not show an increase after this clearance, unlike the other episodes. This may be explained by the deliberate clearance for cereals in a pocket of more base-rich soils. If this was the case, there would be less probability of a nearby population of Calluna which could have readily invaded. The increased occurrence of wetland indicators could confirm this, as more base-rich soils would occur in flushed, and therefore wet areas. They could also indicate that such an area would be susceptible to water-logging after deforestation, which puts the possibility of the success of such a venture in doubt.

Perspective

If the identification of cereals can be accepted then this phase can only be explained in anthropogenic terms. Indeed, it provides one of the clearest representations of pre-Elm Decline 'landnam' type clearances. The implications of this episode are wide-reaching, in particular on the change to the Neolithic and for trade and exchange systems in the Mesolithic, and on the economic sophistication of Mesolithic man. This economic sophistication underlines the strong pressures exerted on an essentially hunter-gatherer mode of existence within a forested and circumscribed environment.

Clearance at SMC 500 cm

2915 ± 50 bc

It is worthwhile repeating at this stage that the occurrence of clearance levels at SMC 500 cm and SMD 500 cm is entirely coincidental and they are not contemporary. SMC 500 occurs only 100 years after the abandonment of the long period of clearance for grazing represented by the second Potentilla peak.

Description

SMC 500 occurs at the period of the maximum extent of deciduous woodland on the plateau. AP values decline, the major loss being to Quercus values which decrease sharply at 500 cm, and stay low for 9 cm before rising again, although former values are never regained. Alnus values decline for this level only. Ulmus values, along with all other deciduous tree values, were increasing up to this level but they decrease at 500 cm, again, never to fully recover. Betula and Pinus rise for this level only, but Tilia values continue to rise. As with all other clearances at Soyland, Corylus values decrease. The decline is quite sharp, and represents the curtailment of the Corylus regeneration after the second Potentilla peak. They increase after 500 cm, but do not regain former values. Again, this clearance phase occurs at the end of a period of high herbaceous values. Gramineae values rise for this level only. Plantago lanceolata, Rumex acetosa/acetosella and Ranunculus acris type occur at this level. Pteridium remains low.

Interpretation and correlations

SMC 500 represents a short-lived local clearance, but apart from this very little further interpretation can be made. The occurrence of Plantago lanceolata is equivocal, on the one hand suggesting clearance for grazing (Behre 1981), on the other, recalling the arable cultivation of SMD 500 cm. Rumex and Ranunculus may be more indicative of pasture areas (eg Vuorela 1970), but neither genus strongly features in the earlier grazing phases.

If a system of rotation of small clearances made by burning was in operation here, as opposed to the more intensive measures employed earlier, then SMC 500 could well have represented an outlier in a rotation whose centre was away from (and therefore not represented in) the mire. However, if SMC 500 represented a second experiment with cereals, similar to SMD 500, then the lack of many of the ruderals and the cereal pollen grains could be explained by the fact that this clearance at SMC was at a further distance from the mire than SMD 500. This is seen in the generally less sharply defined changes in the pollen spectra for SMC 500 cm. One further possibility, that SMC 500 represents a small clearance for vegetable resources (as defined in the first part of this chapter), is improbable. Corylus, one of the major indicators for this type of clearance, has a profile here which excludes this possibility. Firstly, values in the pre-clearance period are the highest they had been for over a millennium, indicating no need for burning. The loss of Corylus values at and after the clearance confirm that vegetable resources (of a 'gatherer' rather than arable variety) were not being encouraged.

Thus, SMC 500 presents an equivocal picture, particularly hard to interpret within the economic scheme as shown at Soyland. The most interesting aspects of this clearance are, firstly, its relationship to the Elm Decline, and, secondly, the post-clearance events. The date of this clearance at 2915 ± 50 bc, and the loss of Ulmus values which had begun to increase just prior to the clearance, may indicate that SMC 500 does in fact mark the Elm Decline proper. However, the decline in Ulmus values at every clearance at Soyland (even where AP values are not generally affected) means that the loss of elm values in themselves cannot be taken as a criterion. While these values never fully rise to the point they had reached in the immediate pre-clearance levels, they do recover to the 4% TP that they have generally maintained through the diagrams from their first rational limit. In point of fact, this confusion means that the placing of the Elm Decline at 470 cm was a fairly arbitrary decision, and may further indicate that the Elm Decline in Britain as a whole, and certainly at Soyland Moor, does not have the almost totemic value that many palynologists endow it with.

This short clearance is difficult to interpret in management terms. However, the immediate development of heathland may in fact provide an explanation, in that SMC 500 cm could have been the beginning of another phase of clearance for grazing, but the onset of podsolisation prevented this from taking place.

The abandonment of the clearance

Of all the clearance phases, SMC 500 most clearly demonstrates how man's activities caused soil degradation and podsolisation during the Mesolithic. In the pre-clearance period, soils had 'recovered' from the long clearance phase for grazing, seen in the low Calluna values and high Corylus values. After SMC 500, Calluna values dramatically increase (from under 5% TP, to 15-37% TP), and remain high thereafter. This tallies with the first increase in *Calluna* values seen at many other British sites. At SMC, this rise, along with the small peaks in the other ericales and *Empetrum* that occur for a few levels above the clearance, represents the spread of heathland over the area cleared. The continued high *Calluna* values indicate that, this time, the threshold soil pH of 4.5 had been crossed, and podsolisation under *Calluna* would have rapidly occurred (Miles 1978, Dimbleby 1961b). For such an ephemeral clearance to have such a great effect, SMC 500 must represent the finishing touch after many millennia of use (or abuse) by man. Previous clearances did allow heathland to invade, but not to such an extent, and only temporarily.

Perspective

This clearance demonstrates the wide-ranging effects of man on the postglacial vegetation development. Firstly, the spread of heathland at SMC appears to be directly correlated to the clearance level. Whether this is also a satisfactory explanation at SMD, and for all those sites with similar *Calluna* increases at this point, is more difficult to judge. At SMD, no trace of this clearance is seen, yet *Calluna* values increase. Unfortunately, time precluded the close sampling of the much longer (in stratigraphic terms, and possibly in dating) SMD 5. Some close sampling was undertaken around the *Quercus* decline at SMD 420 (cf SMC 500), but no evidence of clearance was apparent. It may be that the clearance at SMC 500 triggered a wider-scale reaction than would be supposed from its ephemeral nature, and indeed, wider-scale clearance could have occurred in nearby areas not represented in the mire deposits. Some sites do demonstrate a very similar course of events (eg Glaisdale Moor, in the North York Moors (Simmons and Cundill 1974a), and, to a lesser extent, Bonfield Gill Head in the same area (Simmons and Innes 1981)). The lack of clearance evidence before the major rise in *Calluna* values at many sites could be due to a number of factors: often peat has only just started to accumulate when *Calluna* values rise, and sampling strategies often preclude finding evidence of clearance.

The open nature of the upland woodlands is here seen as a probable result of man's activities. Until SMC 500, oak, alder and elm values were rising, but after SMC 500, the values of these deciduous tree values never recover. Curiously lime values are not affected by the clearance, and continue rising, strongly suggesting that soil degradation, while wide-spread, was not total. This may be explained by the presence of lime in very circumscribed habitats not affected by the clearance activity.

Clearance in the immediate pre-Elm Decline period

Evidence for clearances just before the Elm Decline is widespread in England, particularly in the well-studied North York Moors (see, for example, Simmons and Cundill 1974a). At Soyland Moor, evidence for clearance is diffuse and difficult to interpret at this period. At SMD, AP values are high throughout zone 5 compared to the other three diagrams, apart from the last few levels for the Elm Decline where they drop sharply as herbaceous values increase. The latter half of SMC 5 shows a much more erratic profile of AP values, but they are generally lower than at SMD 5, probably as a result of the clearance at SMC 500. *Corylus* values are generally low but erratic in all four diagrams. At SMC and SMD, herbaceous values remain low, but some of the ruderals (eg Plantago lanceolata) occur more often than previously. The lack of clear cut evidence for clearance at SMC and SMD at this period is probably due to their increasingly regional pollen input as the mire extends (reinforcing the need for small mires to study clearances). The more local pollen input at SMA and SMB allows clearances to be seen in a less diffuse manner, and two possible clearances, one at each site, are depicted. Generally speaking, the higher herbaceous values at SMA and SMB reflects the more open nature of the vegetation around the mire itself when compared to other areas on the plateau and in the nearby valleys, whose pollen dominates SMC and SMD.

Clearances appear to continue around the Soyland mire right up to the Elm Decline, and the evidence tends to point to grazing use rather than for vegetable resources, although the evidence is less clear than in previous periods, and a case can be made for both hypotheses.

SMA 285-280 cm

At this point, *Corylus* and AP values decline while Gramineae values rise. *Potentilla* type and *Rumex acetosa/acetosella* are common, and *Plantago lanceolata* and *Artemisia* also occur. The presence of *Ilex*, *Sambucus* and *Rubus* pollen reflect the edges of this small clearance. After 280 cm, AP and *Corylus* values recover and *Calluna* values rise. This suggests that *Calluna* heathland develops in some areas of the abandoned clearance, but trees are not entirely excluded from regeneration. Again, the rise in *Calluna* may have been the reason for abandoning this clearance after a relatively short period of time. Most of the herb pollen suggests that this clearance was used for grazing, and it is possible that the increasingly acid soils could no longer maintain good quality browse for long periods of time.

The loss of the high *Tilia* values at 292 cm seems unconnected directly with man's activities, unlike the *Tilia* declines seen in Zone VIIb (Turner 1962).

SMB 419 cm

A short clearance phase characterised by loss of AP and *Corylus* values, and an increase in herbaceous values, including *Pteridium*. *Calluna* increases afterwards. Like SMC 500, this phase is not particularly diagnostic and the short pollen profiles make it difficult to correlate with SMA 285-280. On the whole it appears to occur a little earlier (given that the Elm Decline at both sites is contemporary).

The high Cyperaceae values and the constant presence of *Potamogeton* pollen around this level suggest increased water tables at this period. If increasing rainfall were the reason for this, then the evidence should appear in all four diagrams. The peaks in *Potamogeton* values in SMA and SMD may suggest this, but their contemporaneity is not certain, and indeed, these peaks are more likely to reflect the shifting pattern of pools on the mire surface. Thus, the wetland indicators at SMB 419 may reflect locally increased water tables following deforestation. The actual loss of *Sphagnum* values at and after 419 cm may corroborate this, as it is to be expected that any increased rainfall would affect the wetness of the mire, and trigger *Sphagnum* sporing (Conway 1954).

Discussion: The Effect of Man on the Soyland Moor Environment

Evidence for clearance in the pre-Neolithic periods in Britain is now widespread, and clearance seems to have occurred as far back as the Hoxnian interglacial (West 1956, Turner 1970). In the Mesolithic itself, clearance was often thought to have been confined to the later period, when deciduous woodland covered much of the landscape (eg Simmons and Cundill 1974a). However, there is now extensive evidence for clearance in the earlier periods (eg Jones 1977, 1978) York Moors (Jones 1976, 1977 and Jones et al 1979) has evidence for clearance before woodland had properly established itself. The evidence from Soyland Moor, however, is confined to the later Mesolithic. Clearance here occurred in a variety of forms, and seems to have had a profound effect on the vegetation composition.

Most of the evidence at Soyland seems to point to clearance for herbivores. The first phase occurred at a period when woodland was very light. Despite the early date and open nature of the vegetation, this phase seems to represent the most intensive management seen throughout the diagrams. For a period of over 600 years, the landscape of the central Pennines was radically modified by controlled herds. The second phase of grazing was much longer, but may have been restricted to one area of the plateau. It is during this second phase that a clearance of a very different type occurs nearby, pointing to Mesolithic man's knowledge of, and willingness to experiment with, cultivated grasses. The clearances at SMC 500 cm and SMB 419 cm may be further short experiments with cereals, but lack the strongly diagnostic species needed to corroborate this. Certainly it would appear that cereal growing was not a great success, as seen in the increased values of wetland species.

Evidence for Mesolithic habitation and trackways is not clear, but the sporadic occurrence of *Urtica dioica* throughout all the diagrams and the phase of *Sambucus* at SMD 4 may be representative of the nitrogen rich soils which occur around habitations. In the same way, the sporadic occurrence of such ruderals as *Artemisia*, Chenopodiaceae, and, more rarely, *Plantago major/media*, may represent trackways across the plateau. Certainly, the Soyland Moor area was the major crossing route of the central plateau of the millstone grit area of the Pennines in later periods (witness the Roman road across Blackstone Edge which no doubt provided firmer foundations than the mire of Soyland itself. Indeed the name Soyland means 'muddy place'). The large number of Mesolithic sites around Soyland Moor (see Figure 2, page 18), confirms that the area was well-inhabited, and it is therefore likely that they would have been represented in the pollen diagrams.

Thus, clearances for both animal and vegetable resources were maintained on Soyland. A more detailed discussion of the effect this evidence has on the theories of Mesolithic economy and society, and on the mechanics of culture change, will be given in Chapter 8.

The long term effects of these clearances can be seen directly on the vegetation, and indirectly on the soils. It would appear, firstly, that the extent and density of the woodland cover was affected. Jacobi et al (1976) concluded from a study of Atlantic age peats from the central Pennines that Mesolithic man had drastically reduced the forest cover. AP values (excluding Corylus of 27-33% were found between 350-c 500 m, the altitudes at which the majority of Mesolithic sites are found. Below 350 m, AP values ranged between 60-80%. Above 500 m, the values were above 33%. The three-dimensional technique used also showed that the areas around the mires were very sparsely wooded compared to other areas of the plateau. At Soyland, the corresponding period shows AP values of 35-50% TP, which is somewhat higher than those quoted above. However, it is still well below the 60-80% TP found in the areas just below 350 m, and points to very open woodland as compared to densely wooded. At Soyland there are a number of reasons why these low AP values represent the result of man's activities rather than the result of acidic soils and altitude alone:

1. The evidence for repeated sequences of clearances is strong. This would affect the extent of woodland cover directly if trees themselves were removed, and indirectly in the uplands where the resulting exposure factor can delay regeneration (Pears 1977). Also, Johnson and Dunham (1963: 391) state the a priori argument: "if prehistoric peoples could effectively clear forest ... then by the same token they would also be preventing any natural regeneration over an extensive area". However, this becomes a circular argument resting on the assumption that evidence for clearance has been correctly interpreted.

2. If soil acidity and altitude played a decisive role in limiting forest development to merely open woodland, then, it may be expected that they would also retard, and possibly even stop, the establishment and expansion of the more demanding species of Ulmus, Tilia and Fraxinus, which was not the case.

3. The appearance of Acer shows that soil acidity was not a factor which strongly affected the extent of the woodland here, as this species will only thrive on basic soils. If the appearance of Fagus at SMD 500 is not contamination, then the argument is reinforced.

4. Table 3 gives a list of those herb species which have the strongest correlations with man's activities. The number of species present and their occasional abundance at Soyland strongly associates man with the existence of open ground and reduced woodland cover.

Evidence from elsewhere in Britain corroborates the central Pennine findings. In the northern Pennines, clearances which register AP losses were found to occur only between 480-590 m (Turner and Hodgson 1983). The differences in the altitude at which these clearances occur in the two areas may be explained by the entirely different character of their landscapes. Simmons and Cundill (1974b) found that clearance only took place on the high moors of the North York Moors and not on the hill slopes. AP values were not quoted, but the herbaceous values give a similar picture to that of Jacobi et al (1976) for the central Pennines: on the high moors, herbaceous values

TABLE 3 Clearance Indicators Present at Soyland Moor

* Not usually present in British postglacial pollen diagrams until after the Elm Decline (data taken from Godwin 1975a).

Species	Occurrence at Soyland Moor
* *Acer*	Occurs twice
Ulex type	Sporadic
Ranunculus acris type	Common
Caryophyllaceae	Sporadic
Cruciferae	Common
Chenopodiaceae	Abundant
* *Spergula* type	Occurs twice
* *Malva* type	Occurs once
* cf *Ononis*	1 level (2% TP)
* *Trifolium* type	Common
Lotus	Occasional
Potentilla type	Abundant
* *Rubus*	Sporadic
Poterium sanguisorba	Occurs once
Rumex acetosa/acetosella	Abundant
* *R.* cf *crispus*	Occurs once
* *R. obtusifolius* type	2 levels
Melampyrum	Common
Plantago lanceolata	Sporadic
P. major/media	2 levels
* *Jasione montana*	Occurs twice
* *Galium* type	Occasional
* *Valerianella*	Occurs twice
Valeriana dioica	Occasional
Succisa	Common
Compositae types	Abundant
Artemisia	Abundant

are 30-40%, whereas on the hill slopes just below, these values are only 5-10%.

It is possible that the extent of the woodland cover was reduced, not only by specific clearances, but by extensive, low-intensity burning across the plateau top to provide easy mobility and hunting (although, indeed, this was not necessary at some periods when large tracts of land were kept open by grazing pressure). This would have considerably reduced regeneration. This practice would be self-perpetuating, as the more open the woodland, the more often firing would be required to clear the extra amount of undergrowth that would grow due to the added light reaching the ground.

It is also possible that systematic clearances strongly affected the 'natural' succession of vegetation in this area. The early postglacial development has been interpreted in terms of migration and habitat types, and the biotic factor, especially man, has been largely ignored. Smith (1982) makes a strong case for the need to incorporate man in the model, particularly when looking at the timing of vegetation change. At Soyland, it is possible to view the development of virtually every species within the framework of man. The wide range of ruderals has been attributed to man, as has the early expansion of *Calluna*. The long pine maximum and the delaying of expansion of oak can be attributed to the suppression of oak by grazing pressure thereby prolonging the dominance of pine. The fluctuations in the elm and birch pollen curves reflect the different stages of clearances, making it difficult to judge what would have happened to these values if man had not maintained cleared areas.

Hazel is the one species that has long been thought to have a direct association with Mesolithic man (eg Rawitscher 1945). However, as previously discussed, its early arrival in Britain allowed it to greatly expand (Deacon 1974) without the need (or the evidence) to invoke man's activities. At Soyland, the continued existence of high hazel values can easily be attributed to man, but perhaps not in the direct way postulated by most researchers (eg Perring 1965, Jacobi 1978b, Smith 1970). Here, hazel appears to be the ubiquitous side-product of the regeneration of abandoned clearances. Each clearance phase sees the temporary loss of hazel. If hazel was being deliberately 'cultivated', then care would have been taken *not* to clear it. While this does not mean some areas of hazel scrub were not managed for nuts, or coppiced for hurdles for fencing (as such an abundant resource would not have been ignored), it does suggest that the widespread occurrence of hazel was a *secondary* and not a *primary* result of clearances.

Thus, while the postglacial succession at Soyland is recognisably similar to most other British sites, its differences are more attributable to man than the vague 'regional variations' often used to explain local oddities.

In considering the long term effects of clearances, soils must be considered, in particular the possibility of podsol formation. Podsolisation *sensu stricto* is the transferal of iron and aluminium down a soil profile due to acid soil conditions (Pyatt 1970). Loss of bases through leaching usually causes the initial acidity. While podsolisation occurred during previous interglacials, it was an extremely slow process (Tivy 1971), and the occurrence of podsolised soils in Britain during Zones VIIa and VIIb (Dimbleby 1952a, 1955) represents a much earlier degenerative process (Smith 1982), which may be explained as a result of man's activities (eg Dimbleby 1962a). Unreactive parent material, such as the millstone grit, can support a brown earth mull soil under woodland despite the acidity (Tivy 1971). However, the removal of the woodland by man can have a great effect upon this soil system: the loss of shade, the rapid leaching due to downwashing, and lack of root systems (Pearsall 1968), and the further net nutrient loss in grazing or gathering, combine to allow Calluna to invade. The critical factor in the subsequent vegetation development is the soil pH (Miles 1978). Below pH 4.5, Calluna will probably remain as the 'paraclimax' (Gimingham 1972). Soils would then rapidly podsolise due to the strongly acidyfying effects of Calluna itself (Dimbleby 1952a). If the soil pH is above 4.5, Betula is able to invade, and begin to reverse the mor humus under heather to a mull soil while at the same time shading Calluna out (Miles 1978, Dimbleby 1954a). After this, other, more demanding trees and shrubs could also invade.

Calluna was present on and around the mire at Soyland from the very early postglacial, providing a population which could invade cleared areas: with the exception of SMD 500, all clearance phases are marked by a subsequent peak in ericaceous values, particularly Calluna. After the Potentilla type clearances, the subsequent peak of Betula and then Corylus values and the loss of Calluna suggest that the soil acidity remained above the critical pH of 4.5. After SMC 500 the lowered AP values and the high Calluna values remain, suggesting that the acidity had reached its critical threshold, allowing Calluna to dominate and initiate podsolisation. Corylus and Quercus, the two species most adversely affected by the clearance, do not fully recover from what was a very small and short clearance episode. Both recover slightly c 150 years later at the expense of Vaccinium/Erica and Empetrum. These species have similar properties to Calluna, but their effect on soil acidity is not as strong (D J Read pers comm), and thus may have allowed trees to invade. Calluna values remain high, indicating that the recovery was not total.

The Calluna profile at SMD would seem to indicate that localised podsolisation took place here at an earlier period: Calluna values remain above the 5% TP required to indicate local presence throughout. (The lack of a clear inverse relationship with Sphagnum values at SMD may indicate it is less important on the mire surface itself.) If this is the case, then it may explain why most of the later clearance episodes took place away from SMD. The lack of any clear increase in Calluna values after SMD 500 may suggest that this clearance was carried out in an area which had pockets of more base-rich soils.

Thus, a study of the Calluna profiles at Soyland Moor leads to the following conclusions:

1. The Calluna profiles strongly follow the pattern of man's activities, culminating in large expanses of podsolised soils under heathland in the immediate pre-Elm Decline period. Earlier localised podsolisation seems to have occurred earlier at the pollen source area of SMD.

2. The contradictory evidence for the status of the soils in the pollen diagrams (as discussed in Chapter 5) can now be explained by the capacity of the soils to reverse the incipient podsolisation brought about the clearance. Thus some areas in the immediate vicinity of the mire had changing soil pH values during the postglacial.

3. The early appearance of high Calluna values at Soyland Moor (and Robinson's Moss) may be explained as a result of intensive clearance from an early period. This intensive clearance is perhaps confined to densely populated areas, such as the central Pennines.

The clearance hypothesis has much more detailed evidence to back it up than the hypothesis discussed in Chapter 5 of an inverse relationship between Calluna and Sphagnum on the surface of the mire itself. Soil pollen studies undertaken in the southern Pennines (Radley et al 1974) corroborate this: high Calluna values are only associated with Mesolithic flint sites, and are not associated with the onset of peat accumulation. Thus the spread of heathland, in some areas at least, appears to be directly associated with man's activities. Early mire formation may have been a factor in the initial establishment of Calluna at Soyland, but the lack of an early increase in Calluna values at other mires from the same period (eg Ewe Crag Slack site A in the North York Moors (Jones 1978)), means that further explanation is necessary. The lack of evidence for intensive clearance at these sites corroborates the role of man in the early development of Calluna heath.

Mire formation itself may have been caused by man's activities: as early as the 1920's, it was suggested that deforestation was a probable cause of the widespread formation of blanket bog in the highlands (eg Cheetham 1924). The rainfall in these areas was not high enough to revert oak woodland to peat (Dimbleby 1965), and the climatic interpretation for peat initiation (eg Ball 1975) has been based largely on botanical evidence, which has recently been subject to re-interpretation in anthropogenic terms (Moore 1975a). In the central Pennines today, the lower peat margins coincide with the altitudes at which the majority of Mesolithic sites are situated (Tallis 1975), indicating a possible link between man and blanket bog formation in this area at least. There are two major mechanisms postulated for blanket bog formation in oceanic regions:

1. Higher water tables, due to changes in the evaporation: transpiration ratio on non-permeable substrata to the extent that water-logging occurs (Eyre 1968).

2. Pan formation, which occurs in most fully-podsolised soils, creates an impenetrable barrier, causing waterlogging (Pearsall 1968).

Raised water-tables can, of course, be a result of increased rainfall alone. However, deforestation nearly always results in higher water-tables, due partly to the lack of interception of rainfall by the tree foliage, thus reducing evaporation, and partly to the release of the large amount of water locked within the trees themselves. Pan formation, the result of podsolisation, was most likely to have been caused by deforestation, as discussed above.

Over the central Pennines as a whole, both processes may have been involved in blanket bog formation across the plateau. Early peat formation (at and around the BAT) is not associated with podsolised soils (J H Tallis pers. comm.) and may then be the result of higher water-tables via deforestation. However, the later peat formation (ie from mid-VIIa) does occur over podsolised soil profiles. At Soyland itself, it would appear to be the occurrence of podsolisation around SMC between 4160 ± 40 - 2615 ± 50 bc which triggered peat initiation on other areas of the plateau (ie at SMA and subsequently SMB). The highest *Calluna* values at SMA and SMB occur in the pre-peat deposits, and may corroborate this hypothesis. Certainly, this pre-peat evidence helps to solve the chicken-and-egg problem at SMC, indicating that the high *Calluna* values during late SMC 5 represent podsolisation leading to peat formation, rather than an increase in the area of peat around SMC providing new habitats for *Calluna* to invade. The evidence for increased wetness during SMC 4b may then be a result of the intensive land management seen in that area, and explain the more diffuse nature of the wetland indicators at SMD. At SMD 500 cm, there is a direct correlation between clearance and an increased area of wetland habitats.

Repeated clearance seems to have had a catastrophic effect on those British highlands with base-poor parent material, culminating in podsolisation and blanket bog formation. Valleys in upland areas also have evidence for increased rates of silt deposition in Zone VIIa, despite the extent of deciduous woodland at that period which should have helped to minimise riverine deposition (Simmons *et al* 1981, Smith 1982). In lowland sites where rainfall was not high enough to maintain waterlogged deposits, the widespread development of *Calluna* heathland occurs instead, along with evidence for clearance. On the lower greensands of southern Britain, the sites of Oakhanger in Hampshire (Rankine and Dimbleby 1960) and Iping Common in Sussex (Keef *et al* 1965), clearance was followed by *Calluna* heath development. At Addington in Kent (Dimbleby 1963) a brief Mesolithic occupation was not accompanied by clearance, and no such development is seen. At Soyland, the maximum extent of the deciduous woodland occurred after the diagrams first record indications of raised water-tables. If this was a result of climatic change, then it would be expected also to affect the invasion and/or spread of such species as *Tilia* and *Ulmus*, yet values for both species rise after this point.

Testing the Hypothesis

The evidence from Soyland satisfies the basic requirement for any type of clearance hypothesis: more than one episode occurred, suggesting that, for the Late Mesolithic at least, clearance activity was a standard management technique within the economy. This type of evidence is rare, only Bonfield Gill Head, in the North York Moors (Simmons and Innes 1981) showing a similar pattern of repeated clearance. Hopefully, the dearth of sites with such evidence is a sampling strategy factor, not a reality. Large scale studies such as those of Jacobi *et al* (1976), Turner and Hodgson (1983) and Simmons and Cundill (1974b) do provide regional evidence of extensive Mesolithic clearance. The Soyland evidence provides a more detailed study into the mechanisms used.

As to the lines of corroboratory evidence demanded by the burning hypothesis as discussed early in this chapter, the available data is a little more equivocal. Much of the clearance activity can be seen in the uplands and on sandy substrata, but this is largely a sampling phenomenon. Exceptions do occur, with early clearance of lower ground (eg Kildale Hall, North York Moors (Jones 1976) at 168 m OD). Evidence from the northern Pennines indicates clearance on limestone soils also (Turner and Hodgson 1983). At Soyland itself, because burning seems to play a very small role in the clearances, this activity is not confined to winter or summer. Thirdly, the hypothesis stated that the majority of clearances would be small in themselves, with different clearances maintained at different levels of clearance. The evidence at Soyland provides little corroboration of this. The first Potentilla peak in particular points either to numerous fenced-in areas of grazing, or to an extensive area of forest grazing maintained solely by grazing pressure. This period lasted for over 600 years. The second Potentilla peak appears to be less extensive, but much longer, with one short period of regeneration. SMD 500 cm may be part of a network of small-scale clearance for cereal growing, but the evidence, both here and elsewhere, suggests this is unlikely and that it merely represents an experiment. SMC 500 cm and those clearances at SMA and SMB may be part of a larger network, each one representing a different clearance within the succession.

In fact, the Soyland Moor evidence conforms badly with the requirements of the model set out by Simmons (1969) and Mellars (1975). The evidence suggests that burning was not a major tool, and that the envisaged rotation of clearances was not necessarily maintained. Clearances in the Soyland area differed widely in character, ranging from extensive, but intensively managed grazing areas, to small scale arable plots. A greater complexity and higher level of economic sophistication than has been postulated is envisaged. Soyland provides some of the clearest information we have to date of the management techniques that may have been in widespread use during the Mesolithic. The effects such manipulation had on the vegetation and soils appear to be substantial, and can be clearly traced in the diagrams.

CHAPTER 7

THE ELM DECLINE

The Elm Decline has attracted the attention of many palynologists interested in the history of vegetation in north-west Europe and must constitute the most discussed event in the postglacial. Its apparent synchroneity over a wide area has led to monothetic explanations, none of which entirely fit the evidence. These explanations have led to distortion of the data (eg Rowley-Conwy 1982), tortuous hypotheses (eg Tauber 1965), invalid sampling techniques (eg Garbett 1981), and the untestable hypotheses (eg Troels-Smith 1960). An anthropogenic interpretation has often been applied by botanists to the Elm Decline (eg Turner 1962) rather in the manner in which archaeologists pin the label 'ritual' on any inexplicable artefact. Yet the synchroneity of the event is no more suited to anthropogenic explanation than are climatic or disease explanations, and the decline in elm values in areas with little or no archaeological evidence is not explained (eg north-west Scotland (Pears 1977)).

Recently, more researchers are confining their explanations of the event to their own catchment areas (eg Sims 1973, Pennington 1973), rather than remaining baffled by its synchroneity over vast areas. Smith (1970: 91) states that "It may be that the decline of elm is over-emphasised in both the pollen diagrams and the literature." Certainly, recent studies show that the Elm Decline is varied in its manifestations (eg Pennington 1975), suggesting that it represents the end result of a number of processes from soil degradation (eg Rackham 1976) to lopping for fodder (eg Troels-Smith 1960). Some Elm Declines are associated with 'landnam' type clearances (Iversen 1941, quoted in Sims 1973), such as Blea Tarn in the Lake District (Pennington 1965), while others actually record the loss of clearance indicators, such as Cashelkeelty in Connemara (Lynch 1981). All tree species, especially oak and lime, can be documented to both rise and fall in values alongside the loss of elm values. Also, continued dating of the event shows a span of nearly a millenium in Britain alone (including the standard deviations), and tends to merely conform with the evidence provided by Smith and Pilcher (1973) that the decline of a species is more synchronous than its arrival. This span of a millenium also leaves out of account such sites where elm values remain high (eg Thirlmere in the Lake District (Pennington 1965)), or where elm values decline over a long period (eg Fountain's Earth, Nidderdale (Tinsley 1975)), or where two or more Elm Declines seem to occur (eg Featherbed Moss, Derbyshire (Tallis and Switsur 1973)).

The evidence at Soyland Moor, therefore, will be discussed in local terms only. It was hoped that the three-dimensional nature of the Soyland diagrams would provide good evidence for the sequence during the Elm Decline and discussion is centred on the correlation, dating and interpretation of the four profiles of the decline, rather than trying to provide a review of its postulated causes throughout the whole of north-west Europe. However, the hope remained unfulfilled: the four sites at Soyland Moor seem to provide evidence that this event is not unique to the VIIa/VIIb boundary, not synchronous, and that the changes in the profiles of other pollen types associated with the Elm Decline are not identical, despite the small area which was being studied.

Pre-Elm Decline Losses of Elm Values

Note: To avoid confusion, the capitalised term the 'Elm Decline' is used throughout to refer to the VIIa/VIIb boundary only, and not to the decreases in elm values detected in earlier periods.

All the clearance phases at Soyland Moor are accompanied by a decrease in elm values, some more obviously than others. The first Potentilla peak at SMC and SMD records a perceptible loss of elm values, despite the fact that AP values themselves are rising throughout the phase. The clearance at SMD 500 is a classic miniature elm decline, showing a loss of elm values followed by a 'landnam' clearance, even down to the presence of cereal grains. Only its date of 3870 ± 95 bc, and its position prior to the maximum rise in deciduous tree values differ. All other clearances at SMA, SMB and SMC also record losses of elm values.

The miniature elm declines are not unique to Soyland, and can be seen at the clearance phases of Robinson's Moss, Longdendale (Tallis unpublished), Stump Cross, Grassington (Walker 1956) and Tarn Moss, Malham (Pigott and Pigott 1963). In the North York Moors, most clearance phases do not seem to be accompanied by such a loss, but elm values are generally low and such a loss might be difficult to detect. However, Bonfield Gill Head (Simmons and Innes 1981), the only other site with repeated clearance phases, does show a very similar pattern of elm losses with each clearance. It is doubtful whether these decreases in elm values would prove to be statistically significant. However, the 'real' Elm Decline itself is not always picked out by statistical methods (Birks 1974), being an intuitive phenomenon. This may explain the lack of a decline in elm values associated with clearance phases in the northern Pennines (Turner and Hodgson 1983), as statistical methods were used.

Thus, for the central Pennines at least, losses in elm values appear to be always associated with anthropogenic activity. The 'real' Elm Decline may well represent the culmination of a strategy in use for at least three millennia.

Dating and Correlation of the Elm Decline at Soyland Moor

The radiocarbon dating of the Elm Decline at SMC and SMD provided more problems than it solved. On pollen evidence alone, the placing of the boundaries in the Soyland diagrams was not problematic, and it would have been assumed that these were synchronous given the small area studied. Radiocarbon dating appears to show that this was not the case. To fully resolve the contradictory evidence, a full series of dates from all four sites, with at least three dates from each, would be needed, as well as closer sampling at SMD.

The Elm Decline at SMD

The date for the Elm Decline at SMD is 1450 ± 100 bc, well beyond the normal span of dates, and certainly not comparable to SMC. However, it is improper to dismiss radiocarbon dates out of hand merely because they do not conveniently rise to expectation. Indeed, one of the major uses of radiocarbon dating is to provide an absolute standard within the subjective interpretations of available data. All the other dates from Soyland show no inherent difficulties with the dating of the peat, apart from the young date at the base of the SMD profile. This is unfortunate, in that two of the three dates at SMD could therefore be wrong, but the young basal date probably merely reflects the common problem of downwashing of later humic material in the early stages of peat growth (Moore 1975a). Other sites with anomalous dating of the Elm Decline are usually explained away by stratigraphic features. Thus, the RSMTS 1 site in Upper Teesdale (Turner et al 1973), with an Elm Decline date of 1440 ± 90 bc, which is extraordinarily similar to that of SMD, is explained by an hiatus in peat growth detected in the stratigraphy. Thorpe Bulmer in the Durham lowlands (Bartley et al 1976) with a date of 114 ± 60 bc was explained by the very slow peat growth during this period. However, stratigraphic observations do not always correlate with pollen and radiocarbon dating evidence (Tallis and Switsur 1973) and these dates may be correct. At Soyland, however, there were no stratigraphic indications of an hiatus or slow peat growth so this argument cannot be used. Thus, the only reason for rejecting this date is subjective, based on the similarity of the AP and *Corylus* profile of SMC 5 and SMD 5, suggesting that they are contemporary through to the Elm Decline. Oak, alder and hazel curves in particular are very similar. However, closer sampling of this period at SMD, not undertaken because of lack of time, may point to dissimilarities. (The placing of the Elm Decline almost a metre higher up the profile in SMD (at 365 cm) compared to SMC (at 470 cm) does not confirm that the SMD Elm Decline is in fact dated to a much later period than SMC. Faster peat growth at SMD is seen throughout the profile: for example, the placing of the zone 4/5 boundary is at 540 cm in SMC, but 500 cm at SMD, both of which are dated similarly, 4160 ± 40 bc and 3870 ± 95 bc, respectively.)

Thus further dating and sampling is necessary to solve this problem. If the date is rejected then a contemporary date to SMC may be postulated on the pollen evidence. If the date is accepted then there are two possible explanations. Firstly, the Elm Decline is not a synchronous event and has different manifestations over small areas. This is more probable when the very early date of the Elm Decline (3540 ± 140 bc) at Rishworth Moor, a mere 2 km to the south (Bartley 1975), is compared to the date at SMC (2615 ± 50 bc). The Rishworth date is perhaps complicated in that it comes from the base of the peat profile but any contamination here would tend to generate a later date for the event. Research in progress in Wales (P D Moore, in preparation) confirms the possibility that the Elm Decline is not synchronous and has different manifestations in just one basin. Secondly, it is possible that the Elm Decline is represented at SMD 500 cm, making the decrease in elm values at SMD 365 cm an unconnected event to the 'real' Elm Decline. At 3870 ± 95 bc, SMD 500 has an early date for the Elm Decline, but it is nearer the usual time span than 1450 ± 100 bc. The decrease in elm starts at 496 cm, and is followed at 500 cm with a classic 'landnam' type clearance. The difficulty in accepting this as the Elm Decline is its position at the end of the pine maximum, although, as the pine decline is late at Soyland, this feature is not quite as unusual as it first appears. Hipper Sick in Derbyshire (Hicks 1972) has a primary Elm Decline at an even earlier period, at the BAT, so regarding SMD 500 cm as the Elm Decline would not be without precedent. The major difference between the two sites is that elm values do not recover at Hipper Sick, as they do at SMD. However, this does not undermine the hypothesis of an Elm Decline at this point at SMD, as elm values often increase again after the 'real' Elm Decline (Oldfield 1963, Pennington 1975).

The sharp decrease in pine values 8 cm below SMD 500 cm may indicate an hiatus in peat growth, pointing to SMD 500 representing a later period perhaps nearer to the conventional date for the Elm Decline. However, this does not appear to be a tenable explanation as:

1. There is no stratigraphic evidence for it.

2. Pine values still remain high compared to their later decline at SMC 5 and SMD 5.

3. The date of 3870 ± 95 bc, comparable to that at SMC 4/5 boundary, was measured from the levels above the supposed hiatus, and thus points to continuity of peat growth.

Whichever explanation is correct, the Elm Decline at SMD adds to the growing body of evidence that the Elm Decline is not the homogenous event it has been assumed to be.

The Elm Decline at SMC

Prior to radiocarbon dating, the Elm Decline at SMC was paced at the point (SMC 470 cm) where elm values were consistently below 4% TP, and declined to continuous presence only. However, some researchers place the Elm Decline at the point when elm values decline, but show a slight recovery before declining permanently (eg Oldfield 1963, Pennington 1975). If this was applied to SMC, then the Elm Decline could be placed at SMC 500 cm. In itself, it was not a pressing problem, merely confirming that the Elm Decline is a subjectively observed rather than an absolute event. However, problems arise when trying to correlate the Elm Decline at SMC with the other three Soyland diagrams, and nearby sites.

To postulate the position of the Elm Decline at SMC 500 cm correlates well with the data from SMA 2 and SMB 2, where the pre-Elm Decline pollen spectra were similar to SMC before 500 cm (see Chapter 4). Thus the Elm Decline could be seen as a synchronous event across these three diagrams. Also, SMC 500 cm is dated to 2915 ± 50 bc which is within the conventional span of Elm Decline dates (Smith and Pilcher 1973), unlike the date at SMC 470 cm (2615 ± 50 bc), which is rather late. However, when the data from SMD is added to the equation, the position of the Elm Decline at SMC is most likely to be at 470 cm, as the pre-Elm Decline spectra in both diagrams would then be similar. If this is the case, then the earlier correlations with SMA 2 and SMB 2 could be spurious. The differing types of pollen input at the edges of the mire compared to the centre make chronological correlations based on the pollen spectra difficult.

Therefore, if the Elm Decline at SMC is placed at 470 cm, then it is late compared to other British sites, indicating that the Elm Decline is not as synchronous an event as has been thought. However, if it is placed at SMC 500 cm, then the only reason for rejecting the SMD Elm Decline date of 1450 ± 100 bc (ie the good correlation between the whole of SMC 5 and SMD 5) is no longer valid, and this very late date at SMD must be accepted. Thus, wherever the Elm Decline is placed at SMC, the Soyland Moor data undermine the prevailing idea of the synchroneity of the Elm Decline. This conclusion is emphasised by the date of 3540 ± 140 bc for the Elm Decline at nearby Rishworth Moor (Bartley 1975), which is much earlier than any of the possible Elm Decline dates at Soyland.

Comparison of the Pollen Profiles in the Immediate Post-Elm Decline Period

To try to resolve the confusion of correlation, the pollen profiles in the immediate post-Elm Decline period were compared. Data from nearby Rishworth Moor (Bartley 1975) and an unpublished diagram from Soyland Moor by D D Bartley was also added. This latter site is situated a few hundred metres from the deep peat at Soyland, at the head of the River Ryburn valley where peat formation began in the immediate pre-Elm Decline period. To avoid confusion, it is referred to by its local name of Baitings Gate. These correlations are a rough guide only, partly because of the differing pollen sums used at Bartley's sites, and partly because of the differing types of pollen input during this period at the four Soyland sites. Table 4 summarises the data.

Table 4 shows, firstly, that SMD 470 cm is more likely to represent the Elm Decline than SMC 500 cm. Although the pollen profiles at Soyland Moor are by no means identical, they share some common features, particularly in the increasing herbaceous values. SMC 500 cm has no correlations with the Elm Decline at the other sites. Secondly, it confirms that the range of pollen profile changes that can be associated with the Elm Decline is large even over a small area. The four Soyland sites show common features in the decreasing alder and ash values, but oak and hazel values show a varied response. When compared to Rishworth and Baitings Gate, the differences are magnified, both these latter sites showing, for example, an increase in alder values. The only common features are (apart from the *ipso facto* loss of elm) the declining AP values and the rise in herbaceous values. These are not adequate features on which to base a major zone boundary which has been labelled synchronous throughout north-west Europe, and for which monothetic explanations have been sought for decades.

Interpretations of the Elm Decline at Soyland Moor

Even rejecting the SMD date, the available dates for the Elm Decline in the Soyland area have a span of nearly a millennium. The associated changes within the pollen profiles show no close similarity. Thus no monothetic explanation is applicable even to this small area. However, the loss of elm values accompanying all the clearances phases at Soyland points to an anthropogenic interpretation. There are three possible mechanisms by which this latest decrease in elm values may have been achieved.

Firstly, it is possible that the increasingly acid soils and occurrence of widespread podsolisation, due to deforestation, in the immediate pre-Elm Decline period could have prevented regeneration of elm. Elm prefers base-rich soils, although it does occur on some acid

TABLE 4 Comparison of pollen profiles in the immediate post-Elm Decline levels from SMA, SMB, SMC, SMD, Rishworth Moor and Baitings Gate

Pollen values	SMA	SMB	SMC (470 cm)	SMD	Rishworth Moor	Baitings Gate	SMC (500 cm)
Betula	Increasing	As before	As before	Increasing	As before	Increasing	Declining
Pinus	Declining	As before	As before	As before	As before	Declining	As before
Ulmus	Still declining	Still declining	Still declining	Still declining	Remain low	Still declining	Increasing
Quercus	Declining	Declining	Increase to pre-Elm Decline levels	Increase to pre-Elm Decline levels	Increasing	Declining	Increasing
Tilia	As before	Declining	Increase before declining	Declining	Declining	Disappear	As before
Alnus	Declining	Declining	Remain lower than pre-Elm Decline levels	Remain lower than pre-Elm Decline levels	Increases	Increasing	Increasing
Fraxinus	Declining	Declining	Declining	Declining	As before	First appearance	As before

Contd ....

TABLE 4: Continued

Pollen values	SMA	SMB	SMC (470 cm)	SMD	Rishworth Moor	Baitings Gate	SMC (500 cm)
Total AP values	Declining	Declining	Declining	Declining	Slight decline	-	Declining
Corylus	Declining	Increasing	Declining	Increasing	Increases	Increasing	Increasing
Calluna	Increasing	Declining	Increasing	Slight increase	Remain low	Increase to pre-Elm Decline levels	Large increase
Pteridium	Declining	As before	Declining	Declining	First appearance	Increasing	Declining
Total herbaceous values	Increasing	Increasing	Increasing	Increasing	Slight increase	Increasing	Declining

SMC 500 cm is included to allow a comparison between the two possible Elm Decline levels at SMC.

Rishworth Moor data from Bartley (1975).

Baitings Gate data from Bartley (unpublished).

soils (Rackham 1976). Smith and Taylor (1969) and Pennington (1965) have put forward this explanation for the Elm Decline in western Britain. While there is evidence for increasingly acid soils at Soyland, however, it is not known exactly when soil pH would inhibit elm regeneration, so correlation is speculative. Also, using Calluna as an indicator of soil acidity, elm values which decreased during a clearance phase are seen to recover during the period of maximum Calluna values, which does not indicate soil acidity as a critical factor. However, as in the last instance elm values do not recover, podsolisation may have been an added factor.

The second possibility is the use of tree leaves as fodder (Troels-Smith 1960). If elm branches are lopped, then flowering does not occur for seven or eight years afterwards, unlike other trees which will recover in a much shorter period. Thus, decreases in elm values appear to be selective when they are only reflecting the particular characteristics of the tree. It is even possible that elm was directly grazed by the herds who would be able to reach the lower branches and thus retard flowering. This may well be the explanation for some of the clearances, but does not explain how elm declines are seen to occur at SMD 500 cm, representing an arable phase, and also the first Potentilla peak where all AP values actually rise except elm. The loss of elm prior to SMD 500 could possibly be explained as the use of this area for pasture prior to the planting of cereals, or some form of 'mixed economy' where the cultivated areas were fenced off. The first Potentilla peak's loss of elm was ascribed to the possible ringbarking of elm within pockets of more base-rich soil and picked out as such by the presence of elm itself. This has of course been suggested as a possible explanation for the main Elm Decline (eg Nichols 1979) but could equally be applied to any of the small elm declines seen throughout the diagrams.

No single explanation is necessary, all three factors probably coming into play. Soil acidity may have been critical towards the end of the period studied at Soyland, and possibly the loss of lime values which occurs with or just after the Elm Decline may corroborate this. The slight decline in ash values,a species also preferring base-rich soils, may indicate that this, rather than actual clearance activity, which would encourage ash to expand, is the decisive factor. The increasing indications of clearance activity during the immediate pre-Elm Decline period, suggest that clearance also played a role in the loss of elm, either directly by lopping or ringbarking, or indirectly by contributing to soil acidity levels.

It would appear, then, that the Elm Decline at Soyland, and all the preceding small losses in elm values, have an anthropogenic interpretation. However, the actual mechanism for these losses may not be the same in each case. The dating and correlations of the four sites are difficult to interpret, but do not represent a synchronous horizon. The significance of these conclusions in the change from the Mesolithic to the Neolithic is discussed in the next chapter.

To explain the Elm Decline at Soyland in purely parochial terms may ignore a large body of data for other areas, but it seems to provide a more realistic (if still as confused) approach where an event cannot be explained in monothetic terms. Certainly, the previous losses of elm are the result of localised events and there is no overwhelming evidence to suggest that the latest loss is any different. Also, other features of the Soyland diagrams which are common to many sites have been explained in this localised way. The secondary rise of *Corylus* is a good example of this, as at Soyland it appears to be a direct response to the relaxation of grazing pressure in the area. Whether this, or the explanations put forward for the Elm Decline here, can be extrapolated to other areas is immaterial. However, the data from Soyland Moor can be added to the growing body of evidence that the Elm Decline does not represent a unique and synchronous event. Dates for the event occur from the BAT (as at Hipper Sick (Hicks 1971)), or occurring gradually throughout the Atlantic (as at the other Pennine sites of Valley Bog (Chambers 1978) and Fountain's Earth (Tinsley 1975)), or as late as 2615 bc as at SMC. It is possible that the increasing evidence for disparity in dating will allow more complex and realistic explanations to be made.

CHAPTER 8

THE ARCHAEOLOGICAL IMPLICATIONS

The Economy of the Upland Mesolithic

Current hypotheses

The end of the late glacial was bad news for hunters: the long and highly successful relationship between man and reindeer came to an end with the development of woodland. In the place of reindeer, bison and other large herding animals came the red deer, aurochs, elk, roe deer and pig. These animals provided less meat per capita, congregated in smaller herds, if at all, and had plenty of cover in which to hide from the bow and arrow. Given this situation, the evidence we have for economic strategies to optimise the cull should come as no surprise. Mellars (1975, 1976b) provides the rationale for and the mechanism of burning strategies employed by Mesolithic groups to attract herbivores. Simmons (1969) and many others provide the palynological evidence to support anthropogenic interference of the vegetation. The almost ubiquitous presence of charcoal associated with the palynological evidence underlines the use of fire (eg Jacobi et al 1976 and Simmons et al 1981). Lastly, Jarman (1972) points to the archaeological evidence from all of Mesolithic temperate Europe, which indicates some form of ungulate management: bones of red deer and aurochs on archaeological sites show a predominance of young males, which, as females are behaviourally much easier to cull, indicates conscious and selective culling, possible under some form of management. There is some disagreement over which species predominated in the management strategy, Mellars (1975) and Jarman (1972) suggesting red deer, while Evans (1975), Carter (1975) and Grigson (in Simmons et al 1981) suggest that aurochs (Bos primigenius) was more important, at least during the Late Mesolithic. Both species are susceptible to management, and, from the evidence of sites with bones, it would appear that both were utilised in this way (Grigson 1978).

Bad news for the hunter was good news for the gatherer, as the postglacial environment produced greater amounts of edible vegetable food. It would appear to be axiomatic that the warmer the climate, the greater proportion of vegetable food in the diet (eg Lee and DeVore 1968). Apart from one notable exception (ie the extolled hazelnut), the role of plant food has in the past been greatly underestimated (Clarke 1976). The recent recognition of the role of plant foods does not however entirely compensate for the dearth of direct evidence, but it may help in our interpretation of other aspects of Mesolithic culture: Clarke (1976) makes a strong case for re-interpreting the role of microliths as sickles, knives and graters rather than as projectile heads. Site location and settlement patterns may be a function of plant season and habitat, rather than of ungulate movements, although the effects of the prestige of hunting should not be underestimated. The notable exception of the hazel nut constitutes the sum total of our direct knowledge of plant food in the British Mesolithic. Mellars (1976b) and Jacobi (1978b) suggest that a burning strategy similar to that used for ungulates could also be employed to increase and regularise supplies of what may have been a staple resource. The growth of herbaceous species as a side product would also be regularly utilised.

Thus, the current hypotheses for the Mesolithic economy envisage a closely-managed environment. Whether this could be said to constitute domestication depends entirely on how strict a definition one uses. Red deer and aurochs showed no evidence for genetic changes and the hazelnut did not become the filbert, but the notion of ad hoc hunter-gathering can no longer be maintained.

The evidence from Soyland Moor

The location of the Soyland pollen cores on the edge of the most dense area of Mesolithic occupation known in Britain was specifically chosen to further test these hypotheses. The evidence of the diagrams appeared to strongly corroborate the view that man interfered with the postglacial vegetation in order to optimise productivity, at least in the Late Mesolithic, but the nature of this evidence was disconcerting. To begin with, the majority of the evidence did not appear to support the hypothesis of fire as a clearance agent. Charcoal was found only as small and not very numerous wind-blown particles in the peat, which did not increase in frequency at the clearance levels (det. by J H Tallis). The major phases of clearance were interpreted as long periods (ie hundreds of years) of widespread forest grazing, where tree regeneration was suppressed by grazing pressure alone. It was thought possible that the initial density of ungulates was attracted by fire, but either fencing or herdsmen/dog control maintained the density of the herds. This represents a more intensive strategy than burning, requiring greater effort and the co-operation of a larger group of people.

Clearance evidence for plant food was equally disconcerting. SMD 500 represents one of the earliest and most sharply defined phases of cereal production in Britain, at 3780 ± 95 bc, long before the Elm Decline. While the evidence for pre-Elm Decline cereals is growing (Edwards and Hirons 1984), most of it still comes from Ireland, and Soyland Moor is the only upland site with these early cereals. The date firmly places this phase within the Mesolithic (Smith 1974, Switsur and Jacobi 1975). Burning may have been used to clear the ground for cereal production, but again it is not likely. Charcoal is not well represented, and the selective loss of elm values prior to this phase suggests a different type of clearance. The existence of a contemporary clearance for grazing represented at SMC, only 100 m away, also suggests burning was not used, unless the use of fencing is postulated: the ungulates grazing on grass would have been strongly attracted to such a burnt-over area, ruining the crop.

The evidence of hazel at Soyland Moor appears to negate the hypothesis of burning to increase production. Hazel values decline with every clearance phase, suggesting it was either cleared or heavily grazed, or both. However, hazel never disappeared from the area and regenerated strongly after clearances were abandoned. Thus the open nature of the woodland and the regeneration of hazel post-clearance (both a product of anthropogenic activity) meant that no special strategy for encouraging hazel was necessary.

The Soyland Moor evidence as a whole indicates that more intensive economic strategies were practised in this area of the central Pennines than elsewhere. The long period of forest grazing at Robinson's Moss (Tallis unpublished) contemporary with the first Potentilla peak at Soyland suggests that these strategies may have been relatively widespread in the central Pennines. These intensive strategies may be a product of the dense occupation of the area, as a higher population would require greater productivity per unit of land. Equally, however, the initial adoption of more intensive economic strategies would have resulted in denser occupation because of the greater amount of labour input and co-operation required. This is the familiar chicken-and-egg situation of prehistoric culture change processes. In this particular situation, the lack of evidence for similar intensive strategies and density of occupation in other areas at this period may support the latter model.

The Adoption of Agriculture

Invasion or internal development?

Despite the rejection of the invasion model for virtually every other period of prehistory, the explanation for the British Neolithic stubbornly remains in the hands - and oars - of the mainland European farmers: "a first boat arrived from the European mainland containing people who expected to live by farming. It was doubtless followed by others." (Fowler 1983: 3). This subjective view is held even by those willing to treat the Neolithic data in less traditional ways (eg Whittle 1978, Bradley 1978). The evidence rests on the need to import both sheep and cereals, the one not being indigenous to Britain, and the other unlikely to have been domesticated here. This is backed up by the present state of knowledge, firstly for the (dated) end of the Mesolithic at c 3500 bc, (Mellars 1974), while the majority of the dated evidence for the Neolithic begins at 3300 bc (Smith 1974). Secondly, apart from the Irish evidence, there was thought to be no evidence for cereals or 'landnam' episodes in pollen diagrams prior to 3200 bc (ie prior to the Elm Decline). These data appeared to mark an effective break between the two cultures.

However, such negative evidence does not constitute adequate grounds for postulating invasions of peoples. On *a priori* grounds, positive evidence is required before such an hypothesis may be postulated (eg Dennell 1983). In this case, there is evidence to indicate internal assimilation of farming techniques. Firstly, the dating evidence for the end of the Mesolithic is scanty, and a number of researchers agree that it may extend well beyond 3500 bc (eg Barnes 1982, Radley 1969). Secondly, continuity of site location can be seen into the early Neolithic (Bonsall 1978, Morrison 1980, Barnes 1982 and Radley 1969). Thirdly, the evidence for pre-Elm Decline cereals on the British mainland is growing, while the use of the Elm Decline as a preliminary to wide-scale agriculture is by and large discredited (Edwards and Hirons 1984, Jones *et al* 1979). It is left to Dennell (1983) to point out the obvious: that Mesolithic sea-faring is well-documented, while our knowledge of the European farmers indicates that they were "essentially land-based" (Bradley 1978: 99), suggesting that any movement was in the opposite direction to that postulated.

The evidence from Soyland Moor corroborates the theory of trade and exchange and an internal development of agriculture. Early cereal production, at a time when there is no evidence for Neolithic artefacts, and the long-used techniques of herding both point to this.

Explanation for change

The dynamics of culture change remains a vexed question in archaeological methodology. Systems theory provides such a watertight network of negative feedback that a deus-ex-machina is required to allow for culture change (Bender 1978), while the popular catch-all model of population pressure of Boserup (1965) has been disproved on numerous occasions (eg Grigg 1979). It was not part of this research to study the current hypotheses; therefore, as with the Elm Decline, no monothetic explanations are sought for the widespread adoption of agriculture, and only the evidence of Soyland Moor, and other central Pennine studies, is discussed.

The Soyland Moor evidence indicates that herding first occurs in the Late Mesolithic, as represented by the first *Potentilla* peak, starting at c 5959 bc. While this type of forest grazing was more intensive than the more usual cycle of burning, the same effects would be felt upon the social system. The change from the ownership of a dead, quickly-processed and consumed animal to that of a live one requires a commitment of time, and control over live animals requires energy input (Fleming 1972). Both factors engender a more defined territoriality, based on ownership rather than use, and a more complex set of social interactions due to the necessity of co-operation with larger groups. However, territoriality and co-operation are easily assimilated into the hunter-gatherer society (Mellars 1976b). The extra effort required for herding is offset by increased productivity of a given area, but this development need not be the result of adverse conditions, such as population growth or degradation of the environment. Bender (1978) rightly differentiates between greater productivity of a given area, and greater absolute production. Thus, herding activity could merely represent the concentration of a resource by the exploitation of a smaller area. As previously discussed, the available evidence suggests that this was the case in the central Pennines, as compared to the rest of Mesolithic Britain. The reasons for this concentration may lie in the circumscribed nature of this area: this is the narrowest part of the Pennines (c 12 km across above 333 m OD), and it also roughly corresponds to the restricted area of unreactive parent material with sandy soils which were favoured by Mesolithic groups throughout Britain (Jacobi 1978b). This evidence for a strong bias of site selection provides an argument for the intensification of economic strategies in highly circumscribed favourable areas. It has been suggested that this concentration of sites in the central Pennines was merely a direct result of this area being favoured by ungulates (Barnes 1982), but all the available evidence indicates that Mesolithic peoples were well able to control the density of herds, thus suggesting that natural herd density was not a priority when choosing site location.

If aurochs rather than red deer were the preferred herded species in the Late Mesolithic then the adoption of wide scale agriculture after 3200 bc would entail no change at all in this part of the economy. The introduction of sheep (probably as an alternative to red deer) has been explained by Jarman (1972) as a result of Mesolithic forest clearance, as sheep are better adapted to, and more productive in, open landscapes. In the central Pennines, woodland was open above

350 m for much of the Late Mesolithic at least (Jacobi et al 1976), and the Soyland Moor diagrams indicate a further loss of AP values towards the Elm Decline. The introduction of sheep would be quickly assimilated in such an open landscape. It is also interesting to speculate whether the widespread occurrence of *Calluna* heath in the immediate pre-Elm Decline period in the central Pennines was the factor which helped to time the introduction. While *Calluna* heath was present throughout the postglacial at Soyland and Robinson's Moss, it is only during this late period that irreversible and widespread podsolisation (and therefore permanent *Calluna* heath) may have occurred, effectively reducing the carrying capacity of the area. Sheep is a species which will thrive on *Calluna* heath (possibly helped by burning? (Gimingham 1972)), thus allowing different areas of this landscape to be utilised. Thus, the change from hunting to herding could be accomplished with only minor adjustments to the economic and social system, while the change from red deer to sheep would occur as open ground increased and soils degenerated.

Turning to the change from gathering to crop production, if the annual round of the Mesolithic peoples of northern Britain was divided into 'highland' and 'lowland' seasons, then the adoption of cereal production may have entailed a major reorganisation of the society. The major area of cereal production would be the lowlands (it is possible that the increased abundance of wetland species with cereal production at Soyland indicated a lack of success at this higher altitude). Thus crop production would entail an inversion of the annual round by at least some members (presumably female) of the group. If this annual round model is correct, then the necessary reorganisation may be the reason why crop production occurred later than herding. However, it may be that we regard prehistoric cereal production in a biassed fashion. Today it is a carefully tended monoculture, but in the past, particularly at this experimental stage, cereals could have formed merely an interesting addition to other herbaceous crops which were either encouraged or actually planted. (Unless genetic changes occur we cannot detect if a plant has been gathered or sown.) Clark (1976: 461) states that:

> "there is considerable evidence that plant species not only grew together, but were also gathered together, planted together, and cooked together, eaten together and possibly classified together in complex associations and not in pure Linnean species crops and meals."

This type of agriculture could have begun by burning a small area to encourage edible plants, which may then have been weeded of inedible species. Perennials would remain and spread unless overcropped, and annuals were possibly sown. Helbaek (1954) shows that many of our common weeds formed crops in the past (eg *Chenopodium album*). If this was the case then cereals would have merely extended the range of cultivated species. Again, widespread soil degradation would then be the reason for transferring agriculture to the lowlands. Unfortunately, this hypothesis cannot be tested against the palynological data.

It is interesting to note that all the short clearance phases noted in the Soyland diagram are relatively late compared to the long forest grazing phases. If it were possible to attribute all these phases to clearance for vegetable resources (while only SMD 500 has definite evidence for this, all these phases lack the characteristic herb species associated with the earlier grazing), then a shift of economic strategies from animal to vegetable may possibly be detected during the Late Mesolithic.

The implications for the archaeological data

SMD is the only Soyland site which covers the Early Mesolithic period (8300-6700 bc), and no evidence for clearance is found. This may reflect the smaller number of Early Mesolithic sites in the vicinity (although the available data does not always allow discrimination on this point), suggesting that any postulated clearance activity took place away from the pollen source of the mire. Also, the mire was very small at this stage, with birch growing either on it, or around the edges, effectively masking our anything but very local pollen (Janssen 1967). All recorded clearance activity is from the Late Mesolithic. The intensive economic strategies which appear to be represented are not generally consonant with the archaeological data. These economic strategies indicate that the Mesolithic groups remained relatively sedentary for the summer months at least, and possibly all year round, whereas the archaeological evidence appears to indicate that small temporary hunting camps, with a dearth of domestic and maintenance equipment, constitute the majority of the sites. However, some of the archaeological evidence may suggest longer stay sites, in particular the sort of scatters found on, for example, Blackstone Edge. These have been interpreted by Jacobi (1978a) as summer camps, and possibly also indicate some type of co-operative economic strategies (Mellars 1976c). The possible use of some microliths for processing vegetable food (Clarke 1976) could fill part of the tool kit gap. However, there is still little evidence for many processing activities, although our knowledge of tool use is still based to a large extent on inspired guesswork. From a negative point of view, in the Early Mesolithic there is clear evidence of an annual round between the Pennines and the Lincolnshire Edge (Jacobi 1978a), but no evidence is forthcoming for the Late Mesolithic.

Thus, the archaeological evidence does not complement the palynological data very well, as is so often the case from two different lines of research. If the palynological data has been correctly interpreted as large-scale and intensive clearance, then a reassessment of the archaeological data, with particular attention to tool function, may be necessary.

Conclusions

The four pollen diagrams from Soyland Moor produce evidence which may be interpreted as widespread clearance activity, mainly caused by long term, artificially-heightened forest grazing. Jacobi *et al* (1976) concluded that the light woodland cover between 350-500 m on the central Pennines was a result of clearance activity. The Soyland data gives one possible mechanism whereby this was achieved. Man also appears to have changed the natural vegetation development sequence of the postglacial. The extended pine maximum and the delayed expansion of oak may be a direct result of the forest grazing. Most other tree species were affected in one way or another by man's activities, the selective loss of elm with each clearance most noticeably. The types of clearance activity help to undermine the long-held, but unsubstantiated hypothesis of a Neolithic invasion. The Mesolithic and the Neolithic have a different emphasis, but we are not seeing the imposition of one culture upon another.

As is so often the case, this research has posed more questions than it has answered, and the areas for further research are numerous. Similar close sampling above the Elm Decline would help to establish the nature of the local 'Neolithic' activity. Further cores between SMC and SMD, and beyond SMD might help to establish the extent and type of the cereal production clearance, and dating of the Elm Decline at SMA and SMB may help to sort out the confusion. It would be particularly interesting to locate a similarly early peat site in between Soyland Moor and Robinson's Moss, as these two sites are tantalisingly situated on the northern and southern edges of the densest area of Mesolithic occupation. Modern excavation with particular emphasis on the extent of the sites, activity areas and followed by a detailed study of the flints may also help to resolve the continuing difficulties of integrating the archaeological and palynological data: the one suggesting ephemeral occupation which has been interpreted as representative of an unsophisticated culture, the other emphasising complex economic strategies with a concomitant level of social interaction.

REFERENCES

Andersen S T 1967. Tree pollen rain in a mixed deciduous forest in south Jutland. Review of Palaeobotany and Palynology 3: 267-275.

Andersen S T 1970. The relative pollen productivity and pollen representation of north European trees and correction factors for the pollen spectra. Danmarks Geologiske Undersøgelse Series 2, 96: 6-99.

Andersen S T 1973. The differential pollen productivity of trees and its signficance for the interpretation of a pollen diagram from a forested region. In H J B Birks and R G West (eds.) Quaternary Plant Ecology: 109-116. Oxford, Blackwell Scientific.

Atherden M A 1976. Late Quaternary vegetational history of the North York Moors III Fen Bogs. Journal of Biogeography 3: 115-124.

Atherden M A 1979. Late Quaternary vegetational history of the North York Moors VII Pollen diagrams from the eastern central area. Journal of Biogeography 6: 63-83.

Bahn P G 1978. The 'unacceptable' face of the westen European Upper Palaeolithic. Antiquity 52: 183-192.

Ball D F 1975. Processes of soil degradation: a pedological point of view. In J G Evans, S Limrey and H Cleere (eds.) Effect on Man on the Landscape: the Highland Zone: 20-7. CBA Resarch Report no 11 London, Council for British Archaeology.

Barnes B 1982. Man and the Changing Landscape: A Study of the Occupation and Palaeoenvironment in the Central Pennines. Work notes 3. Liverpool, Merseyside County Council/Merseyside County Museums/ University of Liverpool Department of Prehistoric Archeology.

Bartley D D 1964. Pollen analysis of organic deposits in the Halifax region. The Naturalist 880: 77-87.

Bartley D D 1966. Pollen analysis of some lake deposits near Bamburgh, in Northumberland. New Phytologist 65: 141-156.

Bartley D D 1975. Pollen analytical evidence for prehistoric forest clearance in the upland area west of Rishworth, West Yorkshire. New Phytologist 74: 375-381.

Bartley D D, C Chambers and B Hart-Jones 1976. The vegetational history of parts of south and east Durham. New Phytologist 77: 437-468.

Beales P W 1980. The late Devensian and Flandrian vegetational history of Crose Mere, Shropshire. New Phytologist 85: 133-161.

Beckett S C 1979. Pollen influx in peat deposits: values from raised bogs in the Somerset Levels, south west England. New Phytologist 83: 839-847.

Beckett S C 1981. Pollen diagrams from Holderness, north Humberside. Journal of Biogeography 8: 177-198.

Behre K E 1981. The interpretation of anthropogenic indicators in pollen diagrams. Pollen et Spores 23: 225-245.

Bell J N B and J H Tallis 1973. Biological flora of the British Isles: Empetrum nigrum. Journal of Ecology 61: 289-305.

Bender B 1978. Gatherer-hunter to farmer: a social perspective. World Archaeology 10: 204-222.

Binford L R 1968. Post-Pleistocene adaptations. In S R Binford and L R Binford (eds.) New Perspectives in Archaeology: 313-341. Chicago, Aldine.

Birks H J B 1964. Chat Moss, Lancashire. Manchester Literary and Philosophical Society Memoirs and Proceedings 106: 22-45.

Birks H J B 1968. The identification of Betula nana pollen. New Phytologist 67: 309-314.

Birks H J B 1973. Past and Present Vegetation of the Isle of Skye. London, Cambridge University Press.

Birks H J B 1974. Numerical zonations of Flandrian pollen data. New Phytologist 73: 351-358.

Birks H J B and H H Birks 1980. Quaternary Palaeoecology, London, Edward Arnold.

Bonsall C 1975. British antiquity 1974-1975: Palaeolithic/Mesolithic. Archaeological Journal 132: 304-309.

Bonsall C 1978. British antiquity 1977-1978: Palaeolithic/Mesolithic. Archaeological Journal 135: 309-318.

Boserup E 1965. The Conditions of Agricultural Growth. Chicago, Aldine.

Bradley R 1978. The Prehistoric Settlement of Britain. London, Routledge and Kegan Paul.

Bradshaw R H W 1981. Modern pollen representation factors for woods in south east England. Journal of Ecology 69: 45-70.

Brookes D and K W Thomas 1967. The distribution of pollen grains on microscope slides. I The non-randomness of the distribution. Pollen et Spores 9:621-629.

Buckland P C and M J Dolby 1973. Mesolithic and later material from Misterton Carr, Nottinghamshire. An interim report. Transactions of the Thoroton Society 77: 5-34.

Carlisle A and A M F Brown 1968. Biological flora of the British Isles: Pinus sylvestris L. Journal of Ecology 56: 269-307.

Carter H H 1975. Fauna of an area of Mesolithic occupation in the Kennet Valley considered in relation to contemporary eating habits. Berkshire Archaeological Journal 68: 1-3.

Caseldine C J and A D Gordon 1978. Numerical analysis of surface pollen spectra from Bankhead Moss, Fife. New Phytologist 80: 435-453.

Cavers P B and J L Harper 1964. Biological flora of the British Isles: Rumex obtusifolius L and R. crisups L. Journal of Ecology 52: 737-766.

Chambers C 1978. A radiocarbon dated pollen diagram from Valley Bog, on the Moor House National Nature Reserve. New Phytologist 80: 273-280.

Chambers F M 1983. Three radiocarbon dated pollen diagrams from upland peats, north west of Methyr Tydfil, South Wales. Journal of Ecology 71: 475-487.

Cheetham C A 1924. Pennine peat. The Naturalist 1924: 11-15.

Childe V G 1947. Prehistoric Communities of the British Isles. London, Chambers.

Clapham A R, T G Tutin and E F Warburg 1962. Flora of the British Isles. London, Cambridge University Press, second edition.

Clapham A R, T G Tutin and E F Warburg 1981. Excursion Flora of the British Isles. London, Cambridge University Press, third edition.

Clark J G D 1932. The Mesolithic Age in Britain. Cambridge, Cambridge University Press.

Clark J G D 1936. The Mesolithic Settlement of Northern Europe. Cambridge, Cambridge University Press.

Clark J G D 1952. Prehistoric Europe: The Economic Basis. London, Methuen.

Clark J G D 1954. Excavations at Star Carr. Cambridge, Cambridge University Press.

Clark J G D 1978. Neothermal orientations. In P A Mellars (ed.) The Early Postglacial Settlement of Northern Europe: 1-10. London, Duckworth.

Clarke D L 1973. Archaeology: the loss of innocence. Antiquity 47: 6-19.

Clarke D L 1976. Mesolithic Europe: the economic basis. In G de G Sieveking et al (eds.). Problems in Economic and Social Anthropology: 449-482. London, Duckworth.

Conolly A P 1959. Some climatic and edaphic indications from the late glacial flora. Proceedings of the Linnean Society of London 172: 56-62.

Conway V M 1954. Stratigraphy and pollen analysis of southern Pennine blanket peats. Journal of Ecology 42: 117-147.

Courtney T W 1978. A Stone Age Site: Unstone, Derbyshire. Chesterfield, North Derbyshire Archaeological Committee.

Crump W D and C Crossland 1904. Flora of the Parish of Halifax. Halifax, Halifax Scientific Society.

Cundill P R 1977. The distribution, age and formation of blanket peats on the North York Moors. Proceedings of the Northern England Soils Discussion Group 9: 25-29.

Cundill P R 1979. Contemporary pollen spectra on the North York Moors. Journal of Biogeography 6: 127-131.

Danielsen A 1969. Pollen alanytical late Quaternary studies in the Ra district of Østfold, south east Norway. Arbok Universitetetl Bergen Naturvitenskapelig series 14: 7-38.

Davies J 1964. Tranchet axes from West Yorkshire. Bulletin of the Cartwright Archaeological Group, Bradford 9: 68-72.

Deacon J 1974. The location of refugia of Corylus avellana L during the Weichselian glaciation. New Phytologist 73: 1055-1063.

Dennell R 1983. European Economic Prehistory: A New Approach. London, Academic Press.

Dickson J H 1973. Bryophytes of the Pleistocene. London, Cambridge University Press.

Dimbleby G W 1952a. The historical status of moorland in north east Yorkshire. New Phytologist 51: 349-354.

Dimbleby G W 1952b. Soil regeneration on the north east Yorkshire moors. Journal of Ecology 40: 331-341.

Dimbleby G W 1953. Natural regeneration of pine and birch on the heather moors of north east Yorkshire. Forestry 26: 41-52.

Dimbleby G W 1954a. The origin of heathland podsols and their conversion by afforestation. Eighth International Botanical Congress 13: 74-80.

Dimbleby G W 1954b. Pollen analysis as an aid to dating prehistoric monuments. Proceedings of the Prehistoric Society 20: 231-236.

Dimbleby G W 1955. The ecological study of buried soils. Advancement of Science 45: 11-16.

Dimbleby G W 1957. Pollen analysis of terrestrial soils. New Phytologist 56: 12-28.

Dimbleby G W 1961a. Transported material in the soil profile. Journal of Soil Science 12: 12-22.

Dimbleby G W 1961b. The ancient forest of Blackamore. Antiquity 35: 123-128.

Dimbleby G W 1962a. The Development of British Heathlands and their Soils. Oxford, Clarendon Press.

Dimbleby G W 1962b. The ancient forest of Blackamore. Antiquity 36: 136-137.

Dimbleby G W 1963. Pollen analysis of a Mesolithic site at Addington, Kent. Grana Palynologica 4: 140-148.

Dimbleby G W 1965. Postglacial changes in soil profiles. Proceedings of the Royal Society of London, Series B 161: 355-362.

Dimbleby G W 1978. Plants and Archaeology. London, John Baker, second edition.

Edlin H L 1970. Trees, Woods and Man. The New Naturalist no 32. London, Collins, second edition.

Edwards K J 1979a. Palynological and temporal inference in the context of prehistory, with special reference to the evidence from lake and peat deposits. Journal of Archaeological Science 6: 255-270.

Edwards K J 1979b. Environmental impact in the prehistoric period. Scottish Archaeological Forum 9: 27-42.

Edwards K J 1982. Man, space and the woodland edge: speculations on the detection and interpretation of human impact in pollen profiles. In M Bell and S Limbrey (eds.), Archaeological Aspects of Woodland Ecology: 5-22. BAR International Series No 146/Symposia for the Association for Environmental Archaeology No 2. Oxford, BAR.

Edwards K J and K R Hirons 1984. Cereal pollen grains in pre-Elm decline deposits. Implications for the earliest agriculture in Britain and Ireland. Journal of Archaeological Science 11: 71-80.

Erdtman G 1931. The Boreal hazel forests and the theory of pollen statistics. Journal of Ecology 19: 158-163.

Evans P 1975. The intimate relationship: a hypothesis concerning pre-Neolithic land use. In J G Evans, S Limbrey and H Cleere (eds.) The Effect of Man on the Landscape: the Highland Zone: 43-48. CBA Research Report No 11. London, Council for British Archaeology.

Eyre S R 1968. Vegetation and Soils. London, Edward Arnold, second edition.

Faegri K 1951. An unrecognised source of error in pollen analysis. Geologiska Foereningens I Stockholm Foerhandlinger 73: 51-54.

Faegri K 1956. Recent trends in palynology. Botanical Review 22: 639-664.

Faegri K and P Deuse 1960. Size variations in pollen grains with different treatments. Pollen et Spores 2: 293-298.

Faegri K and J Iversen 1975. Textbook of Pollen Analysis. Oxford, Blackwell Scientific, third edition.

Fleming A 1972. The genesis of pastoralism in European prehistory. World Archaeology 4: 179-191.

Fowler P J 1983. The Farming of Prehistoric Britain. London, Cambridge University Press.

Froom F R 1974. A Mesolithic site at Wawcott, Kintbury. Berkshire Archaeological Journal 66: 23-44.

Garbett G G 1981. The elm decline: the depletion of a resource. New Phytologist 88: 573-585.

Gimingham C H 1960. Biological Flora of the British Isles: Calluna vulgaris L. Journal of Ecology 48: 455-483.

Gimingham C H 1972. The Ecology of Heathlands. London, Chapman and Hall.

Gimingham C H 1975. An Introduction to Heathland Ecology. Edinburgh, Oliver and Boyd.

Godwin H 1949. The spreading of the British Flora considered in relation to conditions of the late glacial. Journal of Ecology 37: 140-147.

Godwin H 1960. The history of weeds in Britain. In J L Harper (ed.) The Biology of Weeds: 1-10. Oxford, Blackwell Scientific.

Godwin H 1975a. The Hitory of British Flora. London, Cambridge University Press, second edition.

Godwin H 1975b. The history of the natural forests of Britain: establishment, dominance and destruction. Philosophical Transactions of the Royal Society of London, Series B, 271: 47-67.

Godwin H, R P Suggate and E H Willis 1958. Radiocarbon dating of the eustatic rise in ocean level. Nature 181: 1518.

Godwin H and E H Willis 1964. Cambridge University natural radiocarbon measurements VI. Radiocarbon 6: 116-138.

Grieg J 1982. Past and present lime woods of Europe. In M Bell and S Limbrey (eds.), Archaeological Aspects of Woodland Ecology: 23-55. BAR International Series No 146/Symposia for the Association of Environmental Archaeology No 2. Oxford, BAR.

Grigg D 1979. Esther Boserup's theory of agrarian change: a critical review. Progress in Human Geography 3: 64-84.

Grigson C 1978. The late glacial and early Flandrian ungulates of England and Wales: an interim review. In S Limbrey and J G Evans (eds.), The Effect of Man on the Landscape: the Highland Zone: 46-56. CBA Research Report No 11. London, Council for British Archaeology.

Groenman-Van Waateringe W 1968. The Elm Decline and the first appearance of Plantago major. Vegetatio 15: 292-296.

Goudie A 1977. Environmental Change. Oxford, Clarendon Press.

Hallam J S 1960. The Mesolithic of the Central Pennines. University of Liverpool, unpublished PhD thesis.

Handley W R C 1954. Mull and Mor Formation in Relation to Forest Soils. Forestry Commission Bulletin no 23. London HMSO.

Hansen H P 1949. Pollen contents of moss polsters in relation to forest composition. American Midland Naturalist 42: 473-479.

Heybroek H M 1963. Diseases and lopping for fodder as possible causes of a prehistoric decline in Ulmus. Acta Botanica Neerlandica 12: 1-11.

Hibbert F A and V R Switsur 1976. Radiocarbon dating of Flandrian pollen zones in Wales and northern England. New Phytologist 77: 793-807.

Hibbert F A, V R Switsur and R G West 1971. Radiocarbon dating of Flandrian pollen zones at Red Moss, Lancashire. Proceedings of the Royal Society of London, Series B, 177: 161-176.

Hicks S P 1971. Pollen analytical evidence for the effect of prehistoric agriculture on the vegetation of north Derbyshire. New Phytologist 70: 647-667.

Hicks S P 1972. The impact of man on the East Moor of Derbyshire from Mesolithic times. Archaeological Journal 129: 1-21.

Higgs E S and M R Jarman 1975. Palaeoeconomy. In E S Higgs (ed.) Palaeoeconomy: 1-8. London, Cambridge University Press.

Hodgson J 1974. The Distribution of Pine in the North East of England during the Boreal Period. University of Durham, unpublished MSc dissertation.

Holdgate M W 1955. The vegetation of some British upland fens. Journal of Ecology 43: 389-403.

Hyde H A and D A Williams 1945. The pollen of lime. Nature 155: 457.

Irwin L L 1975. Deer-moose relationships on a burn in north-eastern Minnesota. Journal of Wildlife Management 39: 653-662.

Iversen J 1944. Viscum, Hedera and Ilex as climatic indicators. Geologiska Foereningens I Stockholm Foerhandlinger 66: 463-483.

Iversen J 1949. Influence of prehistoric man on the vegetation.

Danmarks Geologiske Undersøgelse IV, 3, 6: 5-23.

Jacobi R M 1973. Aspects of the Mesolithic age in Britain. In S Kozlowski (ed.), The Mesolithic in Europe: 237-266. Warsaw, Warsaw University Press.

Jacobi R M 1976. Britain inside and outside Mesolithic Europe. Proceedings of the Prehistoric Society 42: 67-84.

Jacobi R M 1978a. Northern England in the eighth millennium bc: an essay. In P A Mellars (ed.), The Early Postglacial Settlement of Northern Europe: 295-332. London, Duckworth.

Jacobi R M 1978b. Population and landscape in Mesolithic lowland Britain. In S Limbrey and J G Evans (eds.), The Effect of Man on the Landscape: the Lowland Zone: 75-85. CBA Research Report no 21. London, Council for British Archaeology.

Jacobi R M, J H Tallis and P A Mellars 1976. The south Pennine Mesolithic and the ecological record. Journal of Archaeological Science 3: 307-320.

Jacobsen G L and R H W Bradshaw 1981. The selection of sites for palaeovegetational studies. Quaternary Research 16: 80-90.

Janssen C R 1959. Alnus as a disturbing factor in pollen diagrams. Medelingen van het Botanisch Museum en Herbarium van de Rijksuniversiteit te Utrecht 153: 55-8.

Janssen C R 1966. Recent pollen spectra from the deciduous and conifer-deciduous forests of north eastern Minnesota: a study in pollen dispersal. Ecology 47: 804-825.

Janssen C R 1967. Stevens Pond: a postglacial pollen diagram from a small typha swamp in northwestern Minnesota, interpreted from pollen indicators and surface samples. Ecological Monographs 37: 145-172.

Janssen C R 1973. Local and regional pollen deposition. In H J B Birks and R G West (eds.), Quaternary Plant Ecology: 31-42. Oxford, Blackwell Scientific.

Jarman M R 1972. European deer economies and the advent of the Neolithic. In E S Higgs (ed.), Papers in Economic Prehistory: 125-148. London, Cambridge University Press.

Johnson G A L and K C Dunham 1963. The Geology of Moor House. Monographs of the Nature Conservancy No 2. London, HMSO.

Jones R L 1976a. Late Quaternary vegetational history of the North York Moors. IV Seamer Carrs. Journal of Biogeography 3: 397-406.

Jones R L 1976b. The activities of Mesolithic man: further palaeobotanical evidence from north east Yorkshire. In D A Davidson and M L Shackley (eds.), Geoarchaeology: 355-367. London, Duckworth.

Jones R L 1977. Late Quaternary vegetational history of the North

York Moors V The Cleveland dales. Journal of Biogeography 4: 353-362.

Jones R L 1978. Late Quaternary vegetational history of the North York Moors VI. The Cleveland Moors. Journal of Biogeography 5: 81-92.

Jones R L, P R Cundill and I G Simmons 1979. Archaeology and palaeobotany on the North York Moors and their environs. Yorkshire Archaeological Journal 51: 15-22.

Keef P A M, J J Wymer and G W Dimbleby 1965. A Mesolithic site on Iping Common, Sussex, England. Proceedings of the Prehistoric Society 31: 85-92.

Kelly J H 1976. The Excavation of Wetton Mill Rock Shelter, Manifold Valley, Staffordshire. Stoke on Trent, City of Stoke on Trent Museum Archaeological Society.

Komarek E V 1966. The metereological basis for fire ecology. Proceedings of the Annual Tall Timbers Fire Ecology Conference 5: 85-125.

Komarek E V 1967. Fire - and the ecology of man. Proceedings of the Annual Tall Timbers Fire Ecology Conference 6: 143-171.

Komarek E V 1969. Fire and animal behaviour. Proceedings of the Annual Tall Timbers Fire Ecology Conference 9: 161-207.

Kozlowski S K 1973. Introduction to the history of Europe in the early Holocene. In S K Kozlowski (ed.), The Mesolithic in Europe: 1-23. Warsaw, Warsaw University Press.

Lamb H H 1977. Climate: Present, Past and Future. Volume 2. Climatic Hitory and the Future. London, Methuen.

Lamb H H, R P W Lewis and A Woodroffe 1966. Atmospheric circulation and the main climatic variables between 8000 and 0 bc: metereological evidence. In J S Sawyer (ed.), World Climate from 8000 to 0 bc: 174-217. London, Royal Metereological Society.

Lee R B and I DeVore 1968. Problems in the study of hunter gatherers. In R B Lee and I DeVore (eds.), Man the Hunter: 3-12 Chicago, Aldine.

Litchi-Federovich S and J C Ritchie 1965. Contemporary pollen spectra in Canada. II The forest-grassland transition in Manitoba. Pollen Spores 7: 63-87.

Lutz H J 1956. The Ecological Effects of Forest Fires in the Interior of Alaska. Technical Bulletin No 1133. Washington, US Department of Agriculture.

Lynch A 1981. Man and Environment in South West Ireland. BAR British Series No 185. Oxford, BAR.

Mabey R 1972. Food for Free. London, Collins.

Mamakowa K 1968. Lille Bukken and Leroy: two pollen diagrams from Western Europe. Arbok Universitet I Bergen Naturvitenskapelig 4: 42-58.

Manby T G 1964. Some Mesolithic sites in the Peak District and Trent basin. Derbyshire Archaeological Journal 83: 10-23.

Mandrioli P, M G Negrini, C Scarini, F Templeri and F Trombetti 1980. Mesoscale transport of *Corylus* pollen grains in winter atmosphere. Grana 19: 227-233.

McVean D N 1956. Ecology of *Alnus glutinosa* (L) Gaertn. III Seedling establishment. Journal of Ecology 44: 195-218.

Mellars P A 1974. The Palaeolithic and Mesolithic. In C Renfrew (ed.), British Prehistory: a New Outline: 41-92. London, Duckworth.

Mellars P A 1975. Ungulate populations, economic patterns and the Mesolithic landscape. In J G Evans, S Limbrey and H Cleere (eds.), The Effect of Man on the Landscape: the Highland Zone: 49-56. CBA Research Report No 11. London, Council for British Archaeology.

Mellars P A 1976a. Settlement patterns and industrial variability in the British Mesolithic. In I M Longworth and G de G Sieveking (eds.), Problems in Economic and Social Archaeology: 375-399. London, Duckworth.

Mellars P A 1976b. Fire ecology, animal populations and man: a study of some ecological relationships in prehistory. Proceedings of the Prehistoric Society 42: 15-46.

Mellars P A 1976c. The appearance of 'Narrow Blade' microlithic industries in Britain: the radiocarbon evidence. In S K Kozlowski (ed.), Les Civilisations du 8e au 5e Millenaire avant notre Ere en Europe: Palaeoenvironment, Structure d'Habitat, Outillages, Economie: 166-174. Paris, Union Internationale des Sciences Prehistoriques et Protohistorique.

Miles J 1978. The influence of trees on soil properties. Institute of Terrestrial Ecology Annual Report 77: 7-11.

Mittre V 1971. Fossil pollen of *Tilia* from the East Anglian Fenland. New Phytologist 70: 693-697.

Moore P D 1968. Human influence upon vegetational history in north Cardiganshire. Nature 217: 1006-1009.

Moore P D 1975a. Origin of blanket mires. Nature 256: 267-269.

Moore P D 1975b. New confidence in ancient pollen data. Nature 256: 166-167.

Moore P D 1976a. How far does pollen travel? Nature 260: 338.

Moore P D 1976b. Disturbed habitats. Nature 260: 750.

Moore P D 1977a. Ancient distribution of lime trees in Britain. Nature 268: 13-14.

Moore P D 1977b. Problems with pollen. Nature 270: 478.

Moore P D 1980. The reconstruction of the late glacial environment: some problems associated with the interpretation of pollen data. In J J Lowe, J M Gray and J E Robinson (eds.), Studies in the Late Glacial of North West Europe: 151-155. Oxford, Pergamon Press.

Moore P D and D J Bellamy 1974. Peatlands. London, Elek Science.

Moore P D and J A Webb 1978. An Illustrated Guide to Pollen Analysis London, Hodder and Stoughton.

Morrison A 1980. Early Man in Britain and Ireland. London, Croon Helm.

Morrison M E S 1959. Evidence and interpretation of 'Landnam' in the north east of Ireland. Botaniska Notiser 112: 185-204.

Nichols H 1967. Vegetational change, shoreline displacement and the human factor in the late Quaternary history of south west Scotland. Transactions of the Royal Society of Edinburgh 67: 145-187.

Oldfield F 1963. Pollen analysis and man's role in the ecological history of the south east Lake District. Geografiska Annaler 45: 23-40.

Oldfield F 1965. Problems of mid postglacial pollen zonation in parts of north west England. Journal of Ecology 53: 247-260.

Oldfield F 1967. Some notes on the implication of pollen analytical evidence. In R W Steel and R Lawton (eds.), Liverpool Essays in Geography: a Jubilee Collection: 107-120. London, Longman's.

Oldfield F 1970. Some aspects of scale and complexity in pollen analytically based palaeoecology. Pollen et Spores 12: 163-171.

O'Sullivan P E 1973. Contemporary pollen studies in a native Scots pine ecosystem. Oikos 24: 143-150.

Pears N 1977. Pollen analysis: a review of some developments and interpretation problems. Scottish Geographical Magazine 93: 32-44.

Pearsall W H 1968. Mountain and Moorlands. Revised by W Pennington. The New Naturalist Series. Glasgow, Collins.

Pennington W 1965. The interpretation of some postglacial vegetation diversities at different Lake District sites. Proceedings of the Royal Society of London, Series B 161: 310-323.

Pennington W 1969. The History of the British Vegetation. London, English Universities Press.

Pennington W 1973. Absolute pollen frequencies in sediments of lakes of different morphometry. In H J Birks and R G West (eds.), Quaternary Plant Ecology: 79-104. Oxford, Blackwell Scientific.

Pennington W 1975. The effect of Neolithic man on the environment in north west England: the use of absolute pollen diagrams. In J G Evans, S Limbrey and H Cleere (eds.), The Effect of Man on the Landscape: the Highland Zone: 74-86. CBA Research Report No 11. London, Council for British Archaeology.

Pennington W 1980. Modern pollen samples from west Greenland and the interpretation of pollen data from the British late glacial. New Phytologist 84: 171-201.

Pennington W and A P Bonny 1970. Absolute pollen diagrams from the British late glacial. Nature 226: 871-872.

Perring F H 1965. The advance and retreat of the British flora. In C G Johnson and L P Smith (eds.), The Biological Significance of Climatic Changes in Britain: 51-60. London, Academic Press.

Petch J A 1924. Early Man in the District of Huddersfield. Huddersfield, Tolson Memorial Publications.

Pigott C D and J P Huntley 1980. Factors controlling the distribution of Tilia cordata at the northern limits of its geographical range. II History in north west England. New Phytologist 84: 145-164.

Pigott C D and M E Pigott 1963. Late glaical and postglacial deposits at Malham, Yorkshire. New Phytologist 62: 317-334.

Pilcher J R and A G Smith 1979. Palaeoecological investigations at Ballynagilly, a Neolithic and Bronze Age settlement in Co. Tyrone, Northern Ireland. Philosophical Transactions of the Royal Society of London, Series B 286: 345-370.

Pitts M W and R M Jacobi 1979. Some aspects of change in flaked stone industries of the Mesolithic and Neolithic in southern Britain. Journal of Archaeological Science 6: 163-177.

Praglowski J 1970. The effects of pre-treatment and the embedding media on the shape of pollen grains. Review of Palaeobotany and Palynology 10: 203-208.

Prentice I C 1978. Modern pollen spectra from lake sediments in Finland and Finnmark, north Norway. Boreas 7: 131-153.

Proudfoot V B 1971. Man's occupance of the soil. In R H Buchanan, E Jones and D McCourt (eds.), Man and his Habitat: Essays presented to Emyr Estyn Evans: 8-33. London, Routledge and Kegan Paul.

Pyatt D G 1970. Soil Groups of Upland Forests. Forestry Commission Forestry Record No 71. London, HMSO.

Rackham O 1976. Trees and Woodland in the British Landscape. London, Dent.

Radley J 1964. Recent prehistoric finds in the Peak District. Derbyshire Archaeological Journal 83: 96-100.

Radley J 1968. A Mesolithic structure at Sheldon. Derbyshire Archaeological Journal 88: 26-37.

Radley J 1969. The Mesolithic period in north-east Yorkshire. Yorkshire Archaeological Journal 42: 314-327.

Radley J and P A Mellars 1964. A Mesolithic structure at Deepcar, Yorkshire, England, and the affinities of its associated flint industries. Proceedings of the Prehistoric Society 30: 1-25.

Radley J, J H Tallis and V R Switsur 1974. The excavation of three 'Narrow Blade' Mesolithic sites in the southern Pennines, England. Proceedings of the Prehistoric Society 40: 1-20.

Raistrick A 1934. Correlation of glacial retreat stages across the Pennines. Proceedings of the Yorkshire Geological Society 22: 199-214.

Ralston C W and G E Hatchell 1971. Effects of prescribed burning on physical properties of soil. Prescribed Burning Symposia Proceedings 1971: 68-85.

Rankine W F and G W Dimbleby 1960. Further investigations at a Mesolithic site at Oakhanger, Selbourne, Hants. Proceedings of the Prehistoric Society 26: 246-263.

Rawitscher F 1945. The hazel period in the postglacial development of forests. Nature 156: 302-303.

Rowley-Conwy P 1982. Forest grazing and clearance in temperate Europe with special refernce to Denmark: an archaeological view. In M Bell and S Limbrey (eds.), Archaeological Aspects of Woodland Ecology: 199-215. BAR International Series No 146/Symposia for the Association of Environmental Archaeology No 2 Oxford, BAR.

Rybnickova E and K Rybnicek 1971. The determination and elimination of local elements in pollen spectra from different sediments. Review of Palaeobotany and Palynology 11: 165-176.

Rymer L 1976. The history and ethnobotany of bracken. Botanical Journal of the Linnean Society 73: 151-176.

Rymer L 1979. Ethnobotany and native distribution of gorse (Ulex europaeus L) in Britain. Environmental Conservation 6: 211-212.

Sagar G R and J L Harper 1964. Biological flora of the British Isles: Plantago major L, P. media L, P. lanceolata L.Journal of Ecology 52: 189-221.

Scaife R G 1982. Late Devensian and early Flandrian vegetation changes in southern England. In M Bell and S Limbrey (eds.), Archaeological Aspects of Woodland Ecology: 57-74. BAR International Series No 146/Symposia for the Association of Environmental Archaeology No 2, Oxford, BAR.

Simmons I G 1964. Pollen diagrams from Dartmoor. New Phytologist 63: 165-180.

Simmons I G 1969. Evidence for vegetation changes associated with Mesolithic man in Britain. In P J Ucko and G W Dimbleby (eds.), The Domestication and Exploitation of Plants and Animals: 113-120. London, Duckworth.

Simmons I G 1975a. The ecological setting of Mesolithic man in the highland zone. In J G Evans, S Limbrey and H Cleere (eds.), The Effect of Man on the Landscape: the Highland Zone: 57-63. CBA Research Report No 11. London, Council for British Archaeology.

Simmons I G 1975b. Towards an ecology of Mesolithic man in the uplands of Great Britain. Journal of Archaeological Science 2: 1-15.

Simmons I G 1981. Culture and environment. In I G Simmons and M J Tooley (eds.), The Environment in British Prehistory: 282-291. London, Duckworth.

Simmons I G, M A Atherden, P R Cundill and R L Jones 1975. Inorganic layers in soligenous mires of the North York Moors. Journal of Biogeography 2: 49-56.

Simmons I G and P R Cundill 1974a. Late Quaternary vegetational history of the North York Moors. I. Pollen analyses of blanket peats. Journal of Biogeography 1: 159-169.

Simmons I G and P R Cundill 1974b. Late Quaternary vegetational history of the North York Moors. II. Pollen analyses of landslip bogs. Journal of Biogeography 1: 253-261.

Simmons I G and G W Dimbleby 1974. The possible role of ivy (Hedera helix L) in the Mesolithic economy of Western Europe. Journal of Archaeological Science 1: 291-296.

Simmons I G, G W Dimbleby and C Grigson 1981. The Mesolithic. In I G Simmons and M J Tooley (eds.), The Environment in British Prehistory: 82-124. London, Duckworth.

Simmons I G and J B Innes 1981. Tree remains in a North York Moors peat profile. Nature 294: 76-78.

Sims R E 1973. The anthropogenic factor in East Anglian vegetational history: an approach using APF techniques. In H J B Birks and R G West (eds.). Quaternary Plant Ecology: 223-236. Oxford, Blackwell Scientific.

Smith A G 1960. The Atlantic-subBoreal transition. Proceedings of the Linnean Society 172: 38-49.

Smith A G 1965. Problems of inertia and threshold related to postglacial habitat changes. Proceedings of the Royal Society of London, Series B 161: 331-342.

Smith A G 1970. The influence of Mesolithic and Neolithic man on

British vegetation: a discussion. In D Walker and R G West (eds.) Studies in the Vegetational History of the British Islės: 81-97. London, Cambridge University Press.

Smith A G and A E P Collins 1971. The stratigraphy, palynology and archaeology of diatomite deposits at Newferry, Co Antrim, Northern Ireland. Ulster Journal of Archaeology 34: 3-25.

Smith A G, C Grigson, G Hillman and M J Tooley 1981. The Neolithic. In I G Simmons and M J Tooley (eds.), The Environment in British Prehistory: 125-209. London, Duckworth.

Smith A G and J R Pilcher 1973. Radiocarbon dates and vegetational history of the British Isles. New Phytologist 72: 903-914.

Smith I F 1974. The Neolithic. In C Renfrew (ed.), British Prehistory: a New Outline: 100-136. London, Duckworth.

Smith R T 1972. A reconsideration of the role of climate in the development of post-Weichselian forest types. In J A Taylor (ed.), Research Papers in Forest Metereology: an Aberystwyth Symposium: 1-19. Aberystwyth, Cambian News.

Smith R T 1982. Towards a more Wholistic View of British Vegetation History. Working Paper No 328. Leeds, School of Geography, University of Leeds.

Smith R T and J A Taylor 1969. The postglacial development of vegetation and soils in northern Cardiganshire. Transactions and Papers of the Institute of British Geographers 48: 75-95.

Smith W G and C E Moss 1903. Geographical distribution of vegetation in Yorkshire. I. The Leeds and Halifax district. Geographical Journal April 1903: 375-401.

Spratt D A, R E Goddard and D R Brown 1976. Mesolithic settlement sites at Upleatham, Cleveland. Yorkshire Archaeological Journal 48: 19-26.

Spratt D A and I G Simmons 1976. Prehistoric activity and environment on the North York Moors. Journal of Archaeological Science 3: 193-210.

Sturdy D A 1975. Some reindeer economies in prehistoric Europe. In E S Higgs (ed.), Palaeoeconomy: 55-97. London, Cambridge University Press.

Switsur V R and R M Jacobi 1975. Radiocaarbon dates for the Pennine Mesolithic. Nature 256: 32-34.

Switsur V R and R M Jacobi 1979. A radiocarbon chronology for the early postglacial stone industries of England and Wales. In R Berger and H E Suess (eds.), Radiocarbon Dating: Proceedings of the Ninth International Conference at Los Angeles and LaJolla 1976: 41-68. Berkeley, University of California Press.

Tallis J H 1964a. The pre-peat vegetation of the southern Pennines.

New Phytologist 63: 363-374.

Tallis J H 1964b. Studies on southern Pennine peats. I. The general pollen record. Journal of Ecology 52: 323-333.

Tallis J H 1964c. Studies on southern Pennine peats. III. The behaviour of Sphagnum. Journal of Ecology 52: 345-353.

Tallis J H 1973. Studies on southern Pennine peats. V. Direct observations on peat erosion and peat hydrology at Featherbed Moss, Derbyshire. Journal of Ecology 61: 1-22.

Tallis J H 1975. Tree remains in southern Pennine peats. Nature 256: 482-484.

Tallis J H and R H Johnson 1980. The dating of landslides in Longdendale, north Derbyshire, using pollen analytical techniques. In R A Cullingford, D A Davidson and J Lewin (eds.), Timescales in Geomorphology: 189-205. London, John Wiley and Sons.

Tallis J H and V R Switsur 1973. Studies on southern Pennine peats. VI. A radiocarbon-dated pollen diagram from Featherbed Moss, Derbyshire. Journal of Ecology 61: 743-751.

Tallis J H and V R Switsur 1983. Forest and moorland in the southern Pennine uplands in the mid-Flandrian period. I Macrofossil evidence of the former forest cover. Journal of Ecology 71: 585-600.

Tauber H 1965. Differential pollen dispersion and the interpretation of pollen diagrams. Danmarks Geologiske Undersøgelse Series II 89: 7-64.

Tauber H 1967a. Differential pollen dispersion and filtration. In E J Cushing and H E Wright (eds.), Quaternary Palaeoecology: 131-141. New Haven, Yale University Press.

Tauber H 1967b. Investigations of the mode of pollen transfer in forest areas. Review of Palaeobotany and Palynology 3: 277-286.

Taylor J A 1965. Climatic change as related to altitudinal thresholds and soil variables. In C G Johnson and L P Smith (eds.), The Biological Significance of Climatic Change in Britain: 37-50. London, Academic Press.

Tinsley H M 1972. A Palynological Study of Changing Woodland Limits on the Nidderdale Moors. University of Leeds, unpublished PhD thesis.

Tinsley H M 1975. The former woodland of the Nidderdale Moors (Yorkshire) and the role of early man in its decline. Journal of Ecology 63: 1-26.

Tinsley H M 1976. Cultural influences on Pennine vegetation with particular reference to north Yorkshire. Transactions of the Institute of British Geographers New Series 1: 310-322.

Tinsley H M and R T Smith 1974. Surface pollen studies across a woodland/moorland transition and their application to the interpretation of fossil pollen diagrams. New Phytologist 73: 547-565.

Teunissen D and H G C M Teunissen-Van Oorschot 1980. The history of the vegetation in south west Connemara, Ireland. Acta Botanica Neerlandica 29: 285-306.

Thomas A S 1960. The trampling animal. Journal of the British Grassland Society 15: 89-93.

Tivy J 1971. Biogeography: a Study of Plants in the Ecosphere. Edinburgh, Oliver and Boyd.

Troels-Smith J 1960. Ivy, mistletoe and elm: climatic indicators -fodder plants. Danmarks Geologiske Undersøgelse Series IV 4: 4-32.

Turner C 1970. The middle Pleistocene deposits at Marks Tey, Essex. Philosophical Transactions of the Royal Society of London Series B 257: 373-440.

Turner J 1962. The Tilia decline: an anthropogenic interpretation. New Phytologist 61: 328-341.

Turner J 1964. The anthropogenic factor in vegetational history. I Tregaron and Whixall Mosses. New Phytologist 63: 73-90.

Turner J 1965. A contribution to the history of forest clearance. Proceedings of the Royal Society of London Series B 161: 343-353.

Turner J 1975. The evidence for land use by prehistoric farming communities: the use of three-dimensional pollen diagrams. In J G Evans, S Limbrey and H Cleere (eds.), The Effect of Man on the Landscape: the Highland Zone: 86-96. CBA Research Report No 11. London, Council for British Archaeology.

Turner J, V P Hewitson, F A Hibbert, K H Lowry and C Chambers 1973. The history of the vegetation of Widdybank Fell and Cow Green reservoir basin, Upper Teesdale. Philosophical Transactions of the Royal Society of London Series B 265: 327-480.

Turner J and J Hodgson 1979. Studies in the vegetational history of the northern Pennines. I Variations in the composition of the early Flandrian forests. Journal of Ecology 67: 629-646.

Turner J and J Hodgson 1981. Studies in the vegetational history of the northern Pennines. II An atypical pollen diagram from Pow Hill, Co. Durham. Journal of Ecology 69: 171-188.

Turner J and J Hodgson 1983. Studies in the vegetational history of the northern Pennines. III Variations in the composition of the mid-Flandrian forests. Journal of Ecology 71: 95-118.

Turner J and A P Kershaw 1973. A late and postglacial pollen diagram from Cranberry Bog, near Beamish, Co. Durham. New Phytologist 72: 915-928.

Vuorela I 1970. The indication of farming in pollen diagrams from southern Finland. Acta Botanica Fennica 87: 1-40.

Vuorela I 1973. Relative pollen rain around cultivated fields. Acta Botanica Fennica 102: 1-27.

Vuorela I 1977. Pollen grains indicating culture in peat, mud and till. Grana 16: 211-214.

Walker D 1956. A site at Stump Cross, near Grassington, Yorkshire, and the age of the Pennine microlithic industry. Proceedings of the Prehistoric Society 22: 23-29.

Walker D and C A Lambert 1955. Boreal deposits at Kirkby Thore, Westmorland. Data for the study of postglacial history XVI. New Phytologist 54: 209-215.

Waton P V 1982. Man's impact on the chalklands: some new pollen evidence. In M Bell and S Limbrey (eds.), Archaeological Aspects of Woodland Ecology: 75-91. BAR International Series No 146/Symposia for the Association of Environmental Archaeology No 2. Oxford, BAR.

Watson G G 1952. Early Man in the District of Halifax. Halifax, Halifax Scientific Society.

Watts W A 1973. Rates of change and stability in vegetation in the perspective of long periods of time. In H J B Birks and R G West (eds.), Quaternary Plant Ecology: 195-206. Oxford, Blackwell Scientific.

Webb III T, R A Lasecki and J C Bernabo 1978. Sensing vegetational patterns with pollen data: choosing the data. Ecology 59: 1151-1163.

West R G 1956. The Quaternary deposits at Hoxne, Suffolk. Philosophical Transactions of the Royal Society of London Series B 239: 265-356.

Whittle A W R 1978. Resources and population in the British Neolithic. Antiquity 204: 34-42.

Woodhead T W 1906. The ecology of woodland plants in the neighbourhood of Huddersfield. Journal of the Linnean Society 37: 333-406.

Woodhead T W and O G E Erdtman 1926. Remains in the peat of the Southern Pennines. The Naturalist 1926: 245-253.

Wray D A, J V Stephens, W N Edwards and C E N Bromehead 1930. The Geology of the Country around Huddersfield and Halifax. Memoirs of the Geological Survey. London, HMSO.

Wright H E and M L Heinselman 1973. Introduction (to the symposium on the ecological role of fire in natural conifer forests of western and northern North America. Quaternary Research 3: 319-328.

Wright H E and H L Patten 1963. The pollen sum. Pollen et Spores 5: 445-450.

Wright J H 1952. Pollen Dispersion of some Forest Trees. North-easterly Forestry Experimental Station Paper No 46.

Wright H H 1953. Pollen dispersion studies: some practical applications. Journal of Forestry 51: 114-118.

Wymer J J 1963. Excavations at Thatcham: final report. Transactions of the Newbury and District Field Club 11 (2): 41-52.

Wymer J J (ed.) 1977. Gazeteer of Mesolithic Sites in England and Wales. CBA Research Report No 20. Norwich and London, Geo Abstracts with the Council for British Archaeology.

APPENDIX I

Preparation of Pollen Slides

Peat samples

1. Removal of one gram of peat from centre of a one-centimetre slice.

2. Peat placed in boiling tube with 10 ml of 10% potassium hydroxide, placed in boiling water bath for 20 minutes, and occasionally stirred.

3. Sieving of broken-down peat mixture through 100 micron aperture sieve washed through with distilled water.

4. Centrifuging of mixture at standard 3,000 rpm for two minutes. Liquid decanted and pellet retained.

5. Washing and centrifuging of pellet twice with distilled water.

6. Addition of 8 ml glacial acetic acid. Centrifuged.

7. Addition of 8 ml acetolysis mixture (0 parts acetic anhydride, 1 part concentrated sulphuric acid). Stirred and placed in boiling water bath for 60 seconds. Immediately centrifuged.

8. Repetition of glacial acetic acid treatment.

9. Resuspension of pellet with distilled water and 3 drops of 100% potassium hydroxide. Centrifuged.

10. Addition of 4 drops aqueous 1% safranine and 3 ml distilled water. Stirred well before centrifugation.

11. Addition of 2 ml of melted glycerol jelly and thoroughly mixed with the remaining pellet.

12. The mixture was spread onto warmed microscope slides (3 per sample) and covered with a large coverslip.

Pre-peat samples

The above method was used for the sandy pre-peat material with the addition of the following procedure between step 5 and 6:

8 ml of 40% hydrofluoric acid added to pellet and placed in boiling water bath for two hours, with occasional stirring. Centrifuged.

Resuspension of pellet in 10% hydrochloric acid and warming of mixture in water bath. Centrifuged.

APPENDIX II

Pollen Types Found at Soyland Moor

Pollen types identified as by Moore and Webb (1978), and Faegri and Iversen (1975), with common names in brackets. Species names follow Clapham, Tutin and Warburg (1981). Where genus only is given, all species of that genus are represented by the pollen type.

Trees

Betula L	:	(birch)
Pinus L	:	(pine)
Ulmus L	:	(elm)
Quercus L	:	(oak)
Tilia L	:	(lime)
Alnus L	:	(alder)
Fraxinus L	:	(ash)
Sorbus aucuparia L	:	(rowan)
Sorbus undiff	:	all other Sorbus species
Acer L	:	(maple)
Populus L	:	(poplars and aspen)
Fagus L	:	(beech)

Shrubs

Coryloid	:	Corylus avellana L (hazel) and Myrica gale L (bog myrtle)
Salix pentandra type	:	S. pentandra L (bay willow) and some grains of S. herbacea L (least willow)
Salix undiff	:	all except above (willow)
Ilex aquifolium L	:	(holly)
Sambucus L	:	(elder)
Crataegus L	:	(hawthorn)
Prunus type	:	(Prunus L and Rosa L (rose)
Ligustrum	:	L. vulgare L (privet)

Dwarf Shrubs

Ericaceae undiff	:	whole family apart from Calluna
Calluna vulgaris (L) Hull	:	(heather)
Empetrum nigrum	:	(crowberry)

Herbs

Gramineae	:	all except cereals (grasses)
Cerealia	:	(cereals)
Cyperaceae	:	all except Rhynchospora alba (sedges)
Rhynchospora type	:	R. alba L only (white beak-sedge)
Caltha type	:	C. palustris L (king cup)
Anemone	:	A. nemorosa L (wood anemone)
Ranunculus acris type	:	Ranunculus L, Clematis L, Pulsatilla Miller, Actea L.
R. cf trichophyllus L	:	(short-leaved water crowfoot)
Thalictrum L	:	(meadow rue)
Nuphar Sm	:	N. lutea (L) Sm (yellow water-lily)

Cruciferae undiff)		
Hornungia type)	:	all Cruciferae species (many species
Sinapis type)	:	each)
Hypericum perforatum type	:	*H perforatum* L, *H. pulchrum* L and *H. tetrapterum* Fries (St John Wort)
Caryophyllaceae	:	all species except next five types
Lychnis type	:	*Lychnis* L. and *Silene* L
Dianthus type	:	*Dianthus* L, *Saponaria* L, some *Silene* L
cf *Dianthus* L	:	mainly *D. deltoides* L (maiden pink)
Spergula type	:	*Spergula* L and *Spergularia* (Pers.) J and C. Presl. (spurrey)
cf *Illecebrum verticillatum* L	:	(illecebrum)
Chenopodiaceae	:	all species
Malva type	:	*Malva* L and *Lavatera* L (mallow)
cf *Geranium* L	:	(cranesbill)
cf *Erodium* L'Her	:	(storksbill)
cf *Genista* L	:	(dyer's greenweed and petty whin)
Ulex type	:	*Ulex* L (furze), *Serothamnus scoparius* L Wimmer ex Koch (broom) (now *Cytisus scoparius* L Link), and *Genista* L
cf *Ononis* L	:	(restharrow)
Trifolium type	:	most *Trifolium* L species and *Medicago* L
Lotus L	:	(birdsfoot trefoil)
Astragalus type	:	*Astragalus* L (milk vetch), *Oxytropis* DC, and remaining *Trifolium* species
Vicia sylvatica type	:	most *Vicia* L and *Lathyrus* L (vetch)
Filipendula Miller	:	(meadow sweet)
Rubus L	:	(bramble)
R. chamaemorus L	:	(cloudberry)
Potentilla type	:	*Potentilla* L (cinquefoil) and *Fragaria vesca* L (wild strawberry)
Geum L	:	(avens)
Poterium sanguisorba L	:	(salad burnet) now *Sanguisorba minor* ssp *minor* scop.
cf *Sedum rosea* (L) Scop	:	(rose-root)
Saxifraga stellaris L	:	(starry saxifrage)
S. granulata type	:	most species not included in others
S. oppositifolia type	:	*S. oppositifolia* L (purple saxifrage) and *S. aizoides* L (yellow mountain saxifrage)
Chrysosplenium L	:	(golden saxifrage)
Drosera rotundifolia type	:	*D. rotundifolia* L and *D. anglica* Hudson (sundew)
Viscum album L	:	(mistletoe)
Hedera helix L	:	(ivy)
Umbelliferae		
cf *Polygonum viviparum*		
Oxyria type	:	*Oxyria digyna* (L) Hill, *Rumex crispus* L, *R. conglomeratus* Murray, *R. sanguinous* L, *R. maritimus* L and *R. pulcher* L (dock)
cf *Oxyria digyna* (L) Hill	:	(mountain sorrel)
Rumex acetosa L	:	(sorrel)
R. acetosella agg	:	(sheep's sorrel)
R. cf *crispus* L	:	(curled dock)

R. obtusifolius type	:	R. obtusifolius L, R. aquaticus L and R. hydrolapatham Hudson
Urtica type	:	Parietaria L (pellitory-of-the wall), Urtica dioica L (stinging nettle) and U. urens L (small nettle)
Urtica dioica	:	(stinging nettle)
Humulus	:	H. lupulus L (hop)
Lysimachia vulgaris type	:	all except L. nemorum L
Centaurium Hill	:	(centuary)
Menyanthes trifoliata L	:	(bog bean)
Polemonium caeruleum L	:	(Jacob's ladder)
Scrophularia type	:	Scrophularia L and Verbascum L
Rhinanthus type	:	Rhinanthus L (yellow rattle), Veronica L (speedwell), Euphrasia L (eye-bright) and Bartsia L
Melampyrum L	:	(cow-wheat)
Pinguicula L	:	(butterwort)
Stachys type	:	Stachys L (woundwort), Lamium L (dead-nettle) Galeopsis L (hemp-nettle), Ajuga L (bugle) and Scutellaria L (skull-cap)
Plantago major	:	(rat-tailed plantain)
P. media L	:	(hoary plantain)
P. lanceolata type	:	mainly P. lanceolata L (ribwort) but can include some grains of P. coronopus L (buck's horn plantain)
P. maritima L	:	(sea plantain)
cf Jasione montana L	:	(sheep's bit)
Galium type	:	Galium L (bedstraw), Asperula L, Sherardia L, and Rubia L (madder)
Valerianella Miller		
Valeriana officinalis L	:	(valerian)
V. dioica L	:	(marsh valerian)
Scabiosa columbaria L	:	(small scabious)
Succisa pratensis Moench	:	(devil's bit scabious)
Compositae Liguliflorae type	:	most species not included in the following
Compositae Tubuliforae types:		
Bidens type		
Anthemis type	:	mostly Achillae L at Soyland (ident. P D Moore)
Carduus L	:	(thistle)
Aster type		
Artemisia L	:	(wormwood)
cf A. norvegica Fr.	:	(norwegian mugwort)
Alisma type	:	Alisma L, Baldellia ranunculoides (L) Parl, Luronium natans (L) Rafin (water plantains)
Potamogeton section Eupotamogeton	:	most species of pondweed, apart from:
P. section Coleogeton	:	P. filiformis Pers, and P. pectinatus L
Typha L	:	(reedmace)
Sparganium L	:	(bar-reed)

Pteridophytes

Lycopodium clavatum L	:	(stag's horn moss)
L. alpinum L	:	(alpine club moss)
Isoetes L	:	(quillwort)
Osmunda regalis L	:	(royal fern)
Polypodiaceae undiff	:	(ferns)
Pteridium aquilinum L Kuhn	:	(bracken)
Athyrium filix-femina (L) Roth	:	(lady fern)
Dryopteris filix-mas type	:	*D. filix-mas* agg, *Cystopteris* Bernh, *Thelypteris* (now *Gymnocarpum* Newman)
D. carthusiana type	:	*D. carthusiana* (Villar) H P Fuchs and *D. cristata* (L) A Grey (buckler fern)
Polypodium vulgare L	:	(polypody)
Ophioglossum L	:	*O. vulgatum* L (adder's tongue)

Mosses

Sphagnum	:	(bog moss)

APPENDIX III

Identification of Less Common Pollen Types

I am indebted to Dr P D Moore of King's College, London for not only helping me to understand the basis of pollen identification and for access to extensive reference material, but also for his expert opinion on the identification of the more difficult pollen types. All the types listed below have been verified by Dr Moore, and by use of reference material.

Betula nana: no grains of *B. nana* were thought to occur in the Soyland diagrams. However, the measurements necessary to distinguish *B. nana* from tree *Betula* (eg Birks 1968), were not reliable due to the swelling of grains in the glycerol jelly mounting media (eg Praglowski 1970, Faegri and Deuse 1960).

Sorbus aucuparia: Distinguished from *Sorbus* type by long, densely-packed striae visible at X 400. Also, smaller and longitudinally thinner.

Crataegus: distinguished from *Sorbus* type by coarser columellae.

Salix pentandra type: smaller than most *Salix* spp. with larger lumina. It is heterobrochate with winding muri which gives it a 'knotted' appearance which may be confused with *S. herbacea*.

Prunus: similar to *Dryas*, but distinguishable by its equatorial bridge across each colpus. Striae often run in broad swirls.

Cerealia: the 4 grains identified as cereals presented problems. Firstly, the large size of the grains was the major reason for placing them within the *Cerealia*, and also for discarding *Glyceria*. As discussed above, swelling can occur in glycerol jelly. This was overcome by measuring 50 *Corylus* grains from the slide with 3 of the large grass grains (SMD 500 cm, slide 2), and calibrating these measurements to the standard 25 microns (Birks 1973). The average figure produced was then applied first to the 4 large grass grains and then to 20 randomly picked small grass grains on the same slide. The results were:

> *Corylus* = 25 microns
> Large grasses = 49.6, 59, 58.5 microns
> (1 at SMD (R) 500 = 56.5 microns)
> Small grasses = 26 microns average
> (range 22.6 - 29 microns)

On this basis, the grains were deemed to be *Cerealia*. Secondly, poor preservation, particularly in the annulus area, meant identification of genus was impossible, although 1 grain was tentatively identified as *Hordeum*.

Rhynchospora: differs from other Cyperaceae in that the grain is spherical, small and with thick walls containing obvious columellae.

Ranunculus cf *trichophyllus*: the type differs from *R. acris* type in the even distribution of the columellae. The specific species was distinguished by type material.

cf Illecebrum verticallatum: separated from Herniaria by type material.

cf Geranium/ cf Erodium: Erodium has smaller clavae, and the columellae on top of the reticulum are fused into rugulae, unlike Geranium.

cf Genista: separated from Ulex type by type material.

cf Ononis: separated from Melilotus by type material. Has a distinctive 'faded' appearance.

cf Sedum rosea: separated from other Sedum spp by type material.

Mercurialis cf annua: differs from M. perennis in that the heads of the pila are fused to form a reticulum. Rare grains of M. perennis also do this, but all 3 grains, scattered through the Soyland diagrams, show this feature, suggesting that M. annua is represented.

Polygonum viviparum: some grains can be distinguished from P. bistorta by the striae which run equatorially. Most P. bistorta striae run longitudinally.

Oxyria cf digyna: much clearer, more distinctive sculpturing, with a 'straight' reticulum. Very long colpi.

Rumex cf crispus: distinguished from other Oxyria types by type material.

Rumex acetosa/acetosella: in the Soyland material, these two types were difficult to distinguish, but PDM placed most of these grains with R. acetosa.

Humulus: separated from Cannabis by shallower pore depth.

Plantago major/media: P. major could occasionally be separated using type material.

cf Jasione montana: separated from Wahlenbergia by greater echinae density.

Artemisia cf norvegica: separated from other Artemisia species by much larger, more prominent echinae.

APPENDIX IV: Soyland Moor Site D (Repeat) Pollen Counts: 503-497 cm
Sample count: 800 grams per level (not including _Sphagnum_) unless otherwise stated

	SMD (R) 503 cm	SMD (R) 502 cm	SMD (R) 501 cm	SMD (R) 500 cm	SMD (R) 499 cm	SMD (R) 497 cm	SMD (R) 495 cm
Betula	4.875	4.875	4.75	4.875	6.64	6.25	4.00
Pinus	4.75	5.0	6.5	3.75	9.02	3.5	6.875
Ulmus	4.125	3.875	3.0	2.125	4.39	4.75	3.625
Quercus	9.00	9.00	13.875	14.5	12.46	11.375	12.25
Tilia	0.125	0.125	0.125	0.875	0.24	0.25	0.25
Alnus	14.375	14.375	20.5	14.25	12.69	16.5	13.75
Fraxinus	0.125	0.125	0.125	0.5	0.12		0.25
Sorbus aucuparia				0.125			
AP TOTAL	37.375%	37.375%	48.875%	41.0%	45.56%	42.625%	41.0%
Corylus	44.125	46.75	44.0	30.0	42.94	45.5	48.625
Salix undiff	0.125		0.5	0.25	0.12		
Sambucus		0.125					
Prunus				0.125			
SHRUB TOTAL	44.25%	46.875%	44.5%	30.375%	43.06%	45.5%	48.625%
Ericales undiff	0.125			0.375		0.125	
Calluna vulgaris	16.00	11.5	4.375	11.25	8.78	9.625	8.0
Gramineae	1.00	0.25	0.25	3.75	0.59	0.5	0.625

Contd ....

APPENDIX IV contd.

	SMD (R) 503 cm	SMD (R) 502 cm	SMD (R) 501 cm	SMD (R) 500 cm	SMD (R) 499 cm	SMD (R) 497 cm	SMD (R) 495 cm
Cerealia				0.125			
Cyperaceae	1.25	2.25	0.75	6.375	1.42	0.875	0.75
Ranunculus acris type				0.125	0.24		0.125
R. cf trichophyllus			0.125	0.125			
Cruciferae				0.125	0.12		
Filipendula		0.125		0.125		0.125	0.125
Potentilla		0.125		0.125		0.125	
Hedera helix		0.125		0.125	0.12		0.125
Oxyria type				0.125	0.12		
Rumex acetosa/acetosella				0.625			
Lysimachia vulgaris						0.125	
Plantago major				0.25			
P. lanceolata type				1.25			
Liguliflorae type				0.125			
Bidens type						0.125	
Polypodiaceae undiff	0.25	0.125	0.25	0.75	0.12	0.25	0.25
Pteridium aquilinum	0.875	0.25	0.625	2.375	0.83	0.25	
Polypodium vulgare	0.25	0.125	0.25	0.625	0.12	0.125	0.375
NAP TOTAL	19.75%	14.875%	6.625%	28.625%	11.39%	12.25%	10.375%
TOTAL	100%	100%	100%	100%	100%	100%	100%
Sphagnum	12.125%		1.375%	16%	7.0%	4.0%	10.0%
					(843 grains)		

APPENDIX V

Radiocarbon Dating at Soyland Moor

By V R Switsur
University of Cambridge

Sites and Stratigraphy

Soyland Moor has the only lengthy pollen diagram from the region of the Pennine hills that has been dated. Soyland Moor is interesting also because it is surrounded by what is probably the greatest density of Mesolithic archaeological sites in the whole of Great Britain (Switsur and Jacobi 1975). These archaeological sites are located on the mineral soil at the base of the blanket peat, which in most localities has a depth of about two metres and are exposed during periods of erosion or constructional activity. Nearby there are areas being studied in connection with sub-fossil tree remains preserved in the peat (Tallis and Switsur 1984). The depth of the peat at Soyland Moor is about seven metres, allowing an excellent resolution both for the pollen spectra and the radiocarbon dates.

The two sites studies here (Soyland Moor sites C and D) are about 30 metres apart and lie at an altitude of some 381 metres. They are situated just north of the A58 main road from Ripponden to Halifax (which is about 9 miles distant) at its junction with the B6138 on the West Yorkshire/Greater Manchester border and overlook the escarpment west of Ripponden. The nearest site that has been studied palynologically is Rishworth Moor which is about one mile to the south east (Bartley 1975). It is thought that the greater depth of the peat here may be a consequence of a confluence of two moorland streams. The stratigraphy was studied by means of 5 cm diameter cores taken by a Russian corer. These were extruded, wrapped in polythene sheeting and stored in the laboratory at about 5 degrees C. The cores provided samples for both pollen analysis and radiocarbon dating.

The details of the stratigraphy given below are confined to the portion of the core that was subsequently used for dating. The depths of the samples were measured from the surface of the peat.

Stratigraphy: Soyland Moor Site C

510 - 450 cms	*Eriophorum vaginatum* peat with some *Sphagnum* remains and occasional stems of *Calluna*.
550 - 510 cms	*Eriophorum vaginatum* peat with occasional Gramineae.
600 - 550 cms	Highly humified *Eriophorum vaginatum* peat.
664 - 600 cms	Brown almost completely humified peat.
690 - 664 cms	Sandy mineral soil.

The results of the pollen analysis of this core are given in Chapter 4.

The stratigraphy of Soyland Moor D site is very similar to Soyland Moor C.

Radiocarbon Dating

Historically there appear to have been two strategies for sampling long peat cores for radiocarbon dating. In one the samples were taken at regular intervals through the depth of the core and the dates used so as to interpolate the ages of significant changes in the pollen diagram. The other was to extract samples at the appropriate changes in the pollen spectra, usually at pollen assemblage zone boundaries, which were thus dated directly, and the sedimentation rates or absolute pollen influx rates were interpolated from these data. The latter strategy was used in this work. In order to obtain a reasonable temporal resolution the smallest amount of peat was taken at each level that would contain sufficient carbon to produce a good radiocarbon date. In the event the samples required multiple counting so as to obtain a small uncertainty of the date.

Clearly, organic material lying in a deposit for many centuries is likely to become contaminated with carbonaceous substances originating at a different time and hence possessing a different concentration of radioactive carbon. Since the measuring equipment cannot distinguish these differences it is important that the sample should contain organic material from only a single age. Techniques are available, or may be devised, for the removal of contamination providing it is either chemically or physically distinct from the substance that is being dated.

Peat is one of the dating materials which suffers from difficulties in such sample preparation. The major contamination sources are threefold. One possibility can arise from the geological situation of the peat bog in the following manner. Rainwater, charged with modern carbon dioxide from the atmosphere through which it falls, is known, in limestone areas, to dissolve ancient carbonate rocks and transport bicarbonates in groundwater into growing peat bogs. Plants that take all their requirements of carbon from the water in which they are submerged, such as Potamogeton species would thus incorporate this mixture of ancient and modern carbon during the course of their photosynthesis since they do not distinguish between them. Therefore, even whilst living they would have an apparent and measurable radiocarbon age. Such 'natural contamination' cannot be removed and one must be aware of this possibility. The hard water effect is not likely to have occurred at Soyland Moor since the surrounding area is one of Millstone Grit.

A second source of contamination arises from the peat itself. The organic materials from the various depths in the peat have different ages. When these decompose they produce substances such as fulvic and humic acids which may be transported as colloids or dissolved in the water moving through the peat. They can become absorbed by or precipitated amongst decomposing plant remains of a different age. This type of contaminant may be removed, or its amount reduced, by gentle chemical action. Usually there will be some loss of the datable material also, since this will obviously contain similar decomposition products. Providing that the whole of the peat is not dissolved it is possible to obtain a sample for dating. Proof that this gives a reliable date is difficult and is normally judged by the internal consistency of a series of samples. In the particular instance of the Soyland Moor sites the samples were washed repeatedly with distilled water and then dilute alkali and substantial quantities were lost to these reagents. Because of the possibility of mixed dates this soluble material was not analysed separately as has been done in some studies but the peat residue was used. The alkaline solutions used in the preparation are liable to absorb modern carbon dioxide from the atmosphere and so incorporate this high radioactivity into the samples. Hence each was acidified and heated to destroy such possible laboratory contamination, before finally washing very thoroughly.

A third and more insidious contamination in peat appears through its mode of growth. Each layer within the peat will at some period have growing into it a number of roots or rootlets from the living plants of subsequent growth strata. These rootlets carry downwards cellular material containing carbon contemporary with the time of its growth. Since the roots can grow great distances into the peat this carbon could be considerably younger than the layer being penetrated. For a peat bog growing at 5 cms per century roots 150 cms long would be contaminating the lower level with carbon that was three thousand years younger. On the death of the plants the roots would decompose and become incorporated in the peat and so indistinguishable from their surroundings. The effect that this would have on the age determination of the lower level would depend both on the age differential between the strata and the percentage of younger material incorporated. This effect would be continuous, each growing layer affecting those below. There appears to be no rational way of allowing for this problem in the dating of peat bogs. It must be remembered therefore, that the radiocarbon date is younger than the true age of the layer sampled by an unknown amount. The difference may be affected by the rate of growth of the peat and its floral composition. It will probably differ between peat bogs growing in different locations and through different time ranges. Internal consistency appears to be the best guide as to the reliability of the relative dates obtained.

Following their pretreatment the radioactivity of the Soyland Moor samples was measured by a sensitive scintillation spectrometer. This necessitated converting each sample into a suitable compound for counting the radioactivity. In this instance the peat was first oxidised in a high pressure oxygen bomb (Switsur 1972) and the combustion products then carefully purified in order to remove substances like water vapour, sulphur oxides, nitrogen oxides and halogens which have the property of 'quenching' the measured radioactivity and so falsifying the results. The pure carbon dioxide extracted from the combustion products and containing all the radiocarbon to be measured, was divided into two portions. One was for the measurement of the stable carbon isotopes ratio in the mass spectrometer in order to make a correction for isotopic fractionation. The second, major portion was converted, first into lithium carbide in a special reactor and then hydrolysed to produce acetylene gas. The acetylene was purified by cryogenic distillation and dried carefully before synthesising the benzene required for the radioactivity measurements. This was weighed carefully and made into a scintillation cocktail for the spectrometer.

Radiocarbon dates for Soyland Moor Site C

Sample Details	Sample Depth cm	Radiocarbon Age. BP	Uncertainty Years	Laboratory Reference
Low *Ulmus* pollen values. Low *Hedera* pollen values. Slight *Tilia* fall.	472-466	4565	50	Q-2385
Local expansion of herb pollen. *Ulmus* declines	503-498	4865	50	Q-2386
Quercus and *Ulmus* pollen values high. *Tilia* and *Fraxinus* curves continuous. *Pinus* pollen values low.	543-538	6110	40	Q-2387
Calluna expansion. *Corylus* values fall.	554-539	6340	50	Q-2388
Expansion of *Corylus*. *Calluna* pollen values low. Clearance phase ends.	585-578	3975	40	Q-2389
First rise of *Alnus* pollen. *Pinus* continues high. Further expansion of herb pollen.	612-607	7640	40	Q-2390
Expansion of *Pinus* and *Quercus*. Fall in *Corylus* pollen. First herb expansion (clearance phase).	638-632	8110	50	Q-2391
Onset of peat formation. Low values of *Pinus* pollen.	664-657	8650	75	Q-2392

Radiocarbon dates for Soyland Moor Site D

Sample Details	Sample Depth cm	Radiocarbon Age. BP	Uncertainty Years	Laboratory Reference
Low *Ulmus* pollen values. Low *Hedera* pollen values.	368-362	3400	100	Q-2393
Sharp clearance phase. Mixed oak forest. 3 cereal grains.	503-497	5820	85	Q-2394
Onset of peat formation.	664-654	8190	70	Q-2395

A batch procedure was used in the sample measurements. Several specimens of the international radiocarbon contemporary standard oxalic acid and samples of inactive carbon to discover the instrumental blank, together with check samples of known age, were included in each group. The radioactivity of each was measured for intervals of 100 minutes and these were repeated until a sufficient precision was obtained for use in the dating calculations.

The spectrometer output was analysed by computer to obtain the radiocarbon disintegration rates and the various laboratory corrections were applied as well as the isotopic fractionation correction to the appropriate samples. From this information the conventional radiocarbon dates were calculated using the Libby half-life of 5568 years for the radiocarbon isotope and based on the zero year of AD 1950. By international agreement these dates are designated as BP. No attempt has been made in the tables to calibrate these conventional radiocarbon dates to calendar years since most of the earlier dates are beyond the range of even the most recent calibration curve, and calendar years thus have no basis, as yet, for comparison with other dated pollen diagrams.

The dates for the Soyland Moor D site were obtained for a comparison with the main site C. It appears that the onset of peat formation at site D was much later than at site C which could be explained in terms of topography. More likely, though, is the possibility that the early peat formation at site D may have been dispersed by flash flooding or wind erosion.

Peat Accumulation Rates

Figure 6 shows a graph of the radiocarbon ages versus the depth in the core, often termed the sedimentation curve or sometimes the accumulation curve for the peat. A least-squares regression fit through all the points produces a straight line with a regression coefficient of 99.6% and suggesting a mean growth rate of 4.54 cms per radiocarbon century during about four thousand years.

However, a closer inspection of the curve shows that in this regression not all the points fall within one standard deviation of the fitted line. By making use for the small values of the uncertainties of the radiocarbon dates a better idea of variations of the growth rate may be obtained. The accumulation curve may be divided by inspection into four sections, identified in the figure by A, B, C and D. The growth rate falls slightly in period B compared with that of A. During period C a point of inflexion occurs and the rate increases rapidly up to period D. If these shorter periods are considered as linear, regression lines may be fitted to them and the mean accumulation rate obtained for each period. The rates are presented in the table below.

It is interesting that the linear fit incorporating all of the points of the graph leads to an age of 270 years for the uppermost peat layer, indicating that an increase in growth rate over the regression value is required during the later stages of growth to account for this. A loss of peat from the surface might also be argued.

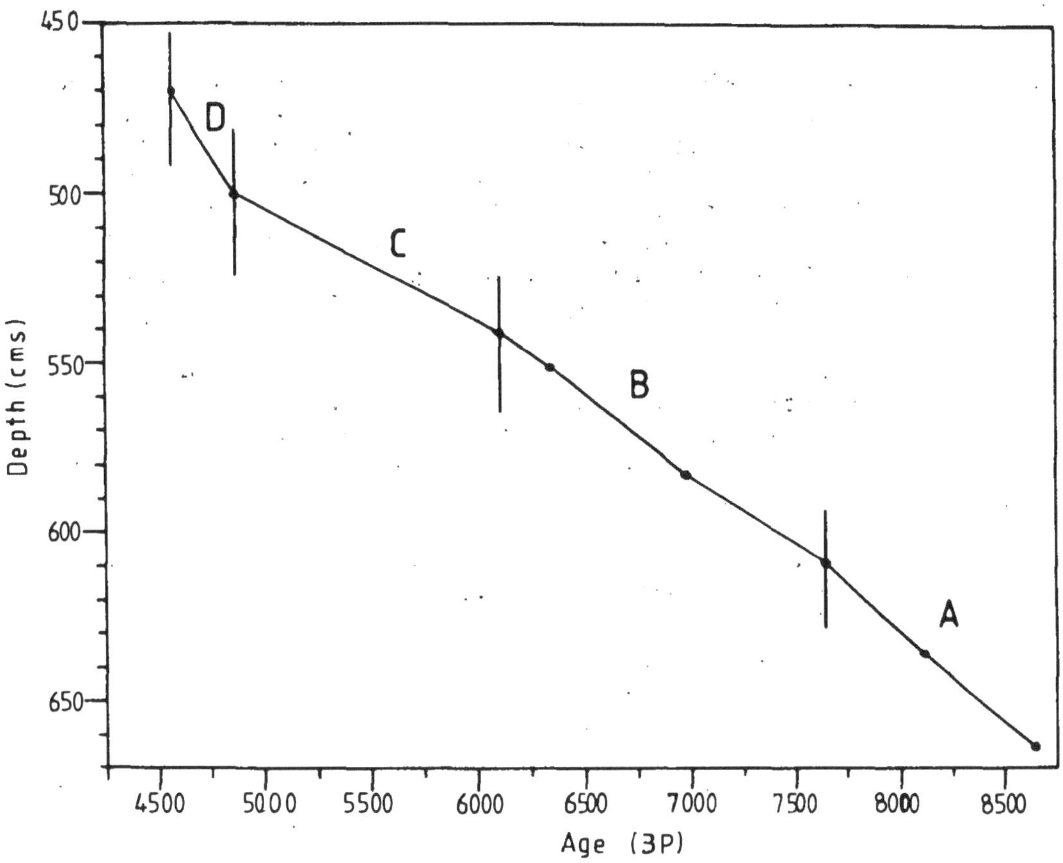

Figure 6 Peat accumulation curve: conventional radiocarbon dates

<u>Variation of the growth rate of Soyland Moor C</u>

Period	Range (BP)	Accumulation rate
A	8650 - 7640	5.4 cm per century
B	7640 - 6110	4.5 " " "
C	6110 - 4860	3.3 " " "
D	4860 - 4565	10.0 " " "

However, if the composite curve is accepted and the portion D extrapolated to zero depth, the age predicted for the surface layer is only about 10 years which, within the uncertainties involved, agrees well and tends to give some support to this approach. Nevertheless it must be pointed out that both of the above scenarios for the accumulation rate require further modification. This is because the basic assumptions concerning the linearity of both the depth and the time axes are not justified.

The peat deposit, as it becomes older, undergoes progressive decomposition and humification and is compacted by the ever increasing overburden of the bog as it continues to grow. Hence the lower levels of the peat become more and more compressed than those overlying them and these in turn are less compressed than the uppermost layers. The mathematical form of this compression depends on the physical characteristics of the decomposing mire and will no doubt be different for different communities but is likely to be of exponential form or follow a power law. The values for the best fit need to be determined in order to discover the original position of the uncompressed peat layer. This would become very difficult in practise and all that can be measured is the apparent accumulation rate of the peat in the present, compacted, situation. How serious the discrepancy might be would depend upon the amount of compaction involved.

The temporal axis of the accumulation graph is also distorted, for it is well known that the constancy of the atmospheric radiocarbon concentration, which is an essential assumption for the simple theory of radiocarbon dating, is only approximate and that in consequence the radiocarbon years are of differing length at different periods. Prior to about 2500 BP the radiocarbon years are foreshortened with respect to solar years in a progressive manner until at about 5000 to 6000 BP. This has the effect of making all radiocarbon dates in this period too young by varying amounts. Thereafter, the discrepancy is believe to decrease, also progressively, until at around 9000 to 10000 years ago the scales once more coincide. This means that all radiocarbon dates have to be calibrated if the true calendrical equivalent is required for purposes such as absolute accumulation rates which would be appropriate here. The calibration of the radiocarbon time scale has

Figure 7 Calibration diagram for radiocarbon dates

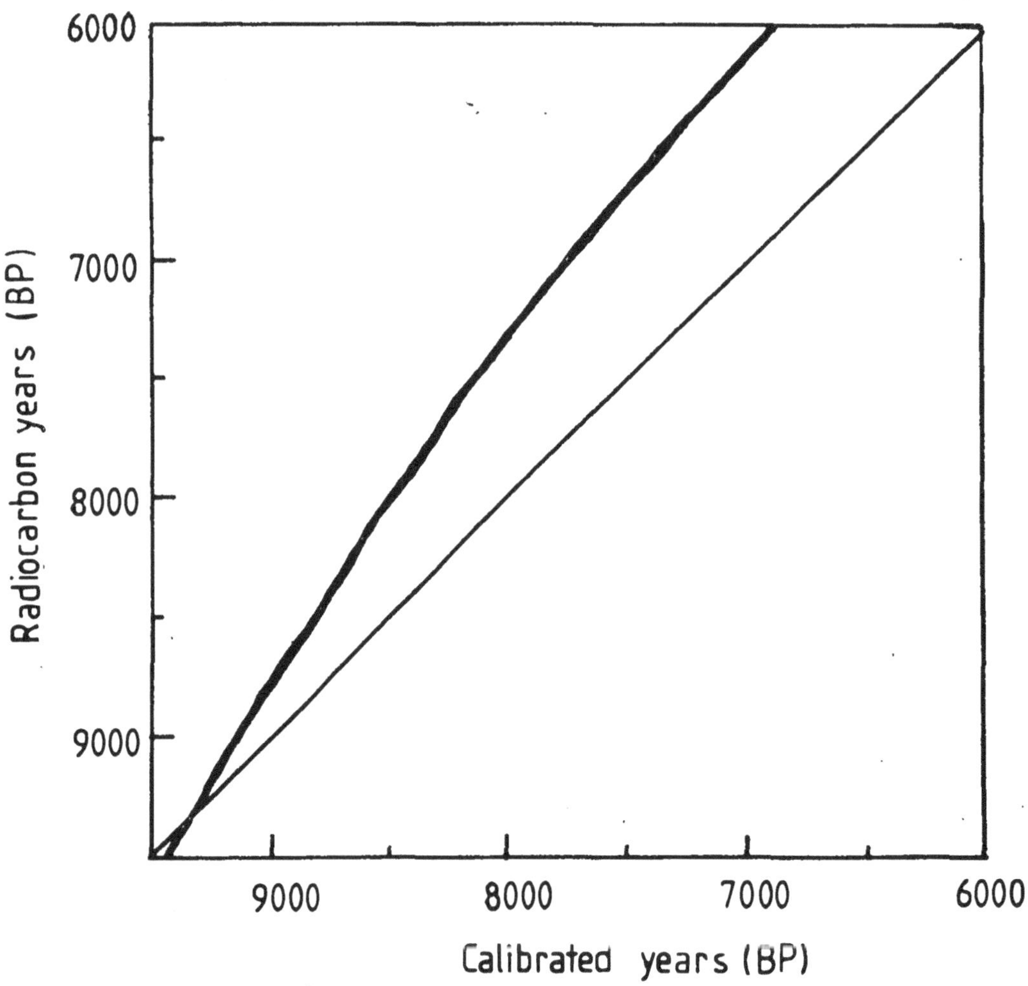

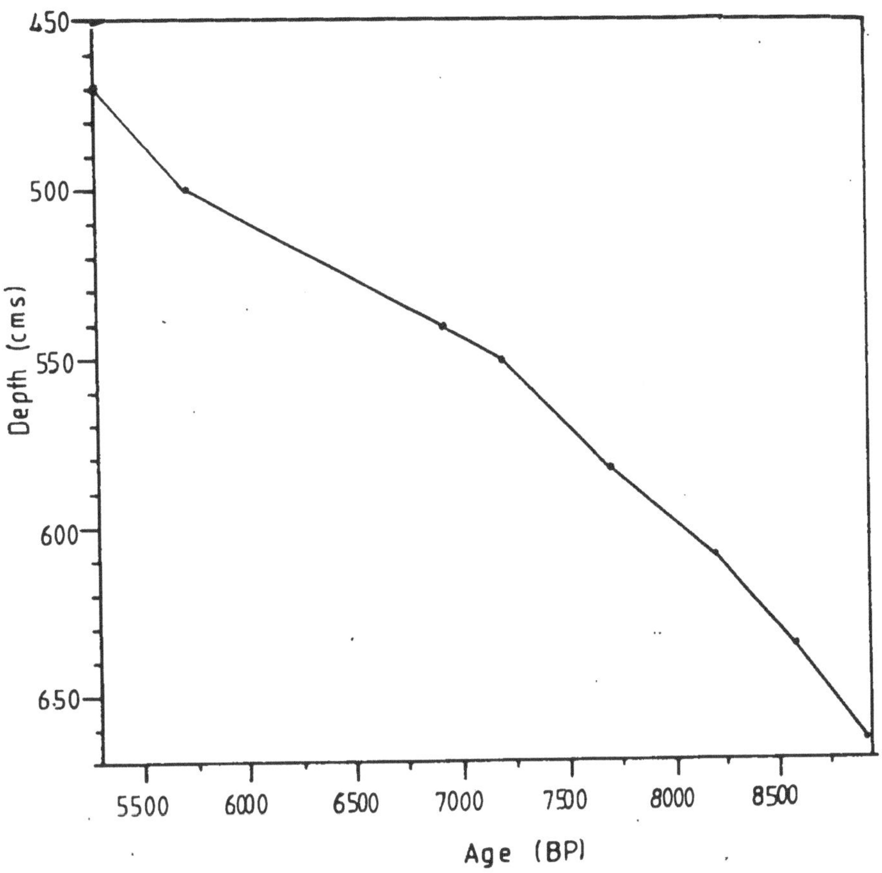

Figure 8 Peat accumulation curve: 'calibrated' dates

been proceeding for nearly a quarter of a century and the newer measurements appear to be precise to about ± 20 years. Unfortunately they reach back in time only some 6000 years so that the majority of the Soyland C accumulation curve is beyond the range. It is possible, nevertheless, with some reasonable assumptions to extrapolate the calibration further back in time, but with decreased precision, and obviously some speculation. Until an authentic calibration curve is established for the period, a demonstration of the effect of this on the present study is suggested. The notional calibration curve shown in Figure 7, which may be precise to perhaps a century or so has been used to give a 'first order' calibration to the conventional radiocarbon dates. The resulting accumulation curve is given in Figure 8. The points lie on a fairly smooth curve indicating a continuously decreasing accumulation rate which may be equated with amelioration of the climate and the gradual onset of dryer conditions.

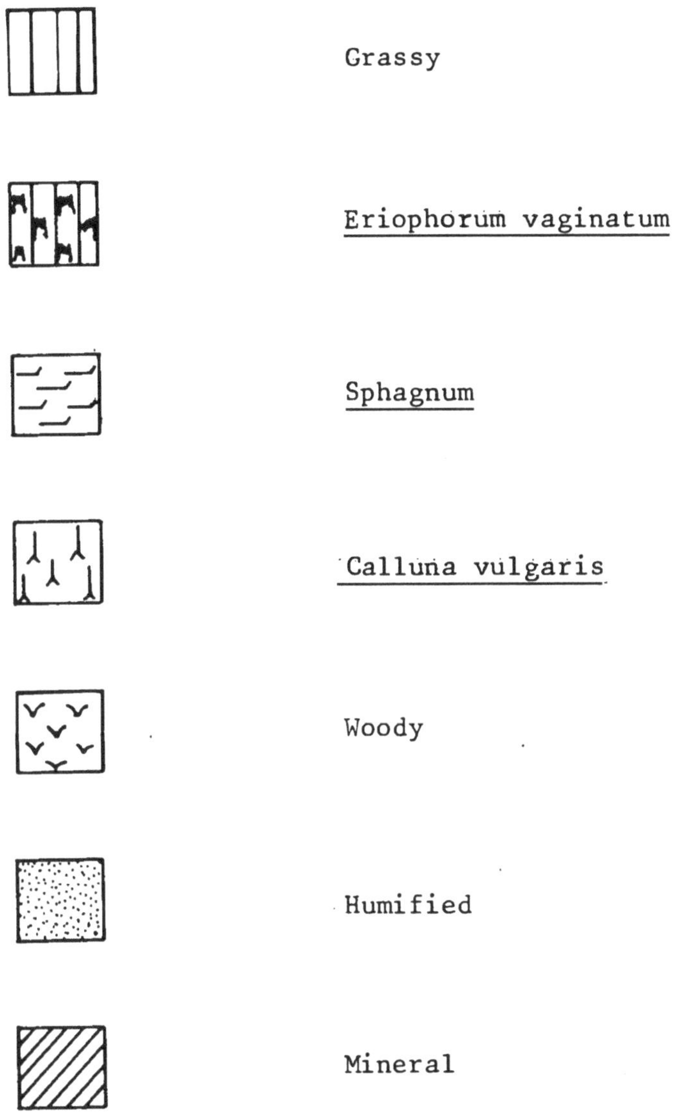

Figure 9 Key: stratigraphic symbols of the pollen diagrams

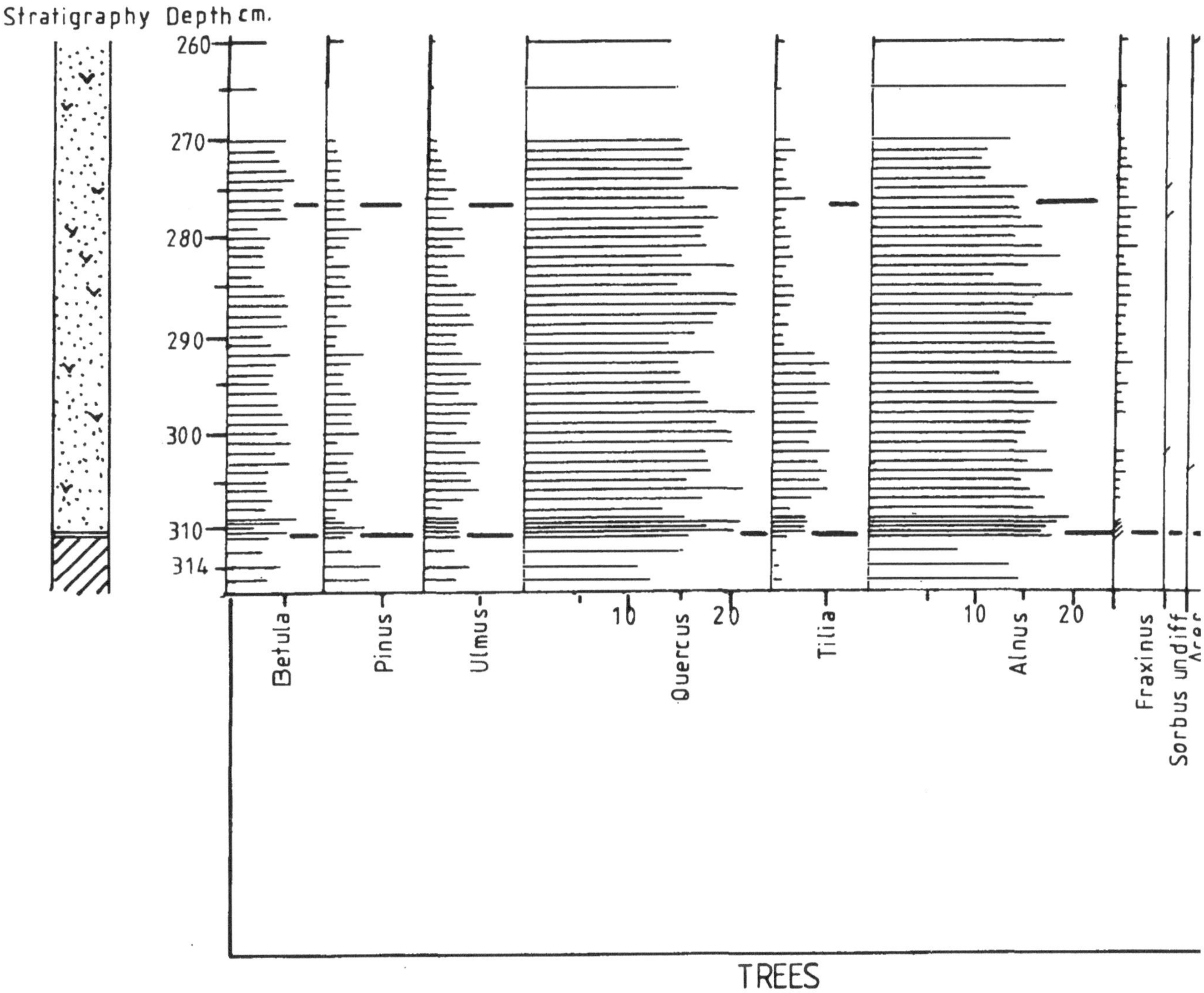
Stratigraphy
Depth cm.
260
270
280
290
300
310
314
Betula
Pinus
Ulmus
10
20
Quercus
Tilia
10
20
Alnus
Fraxinus
Sorbus undiff
TREES

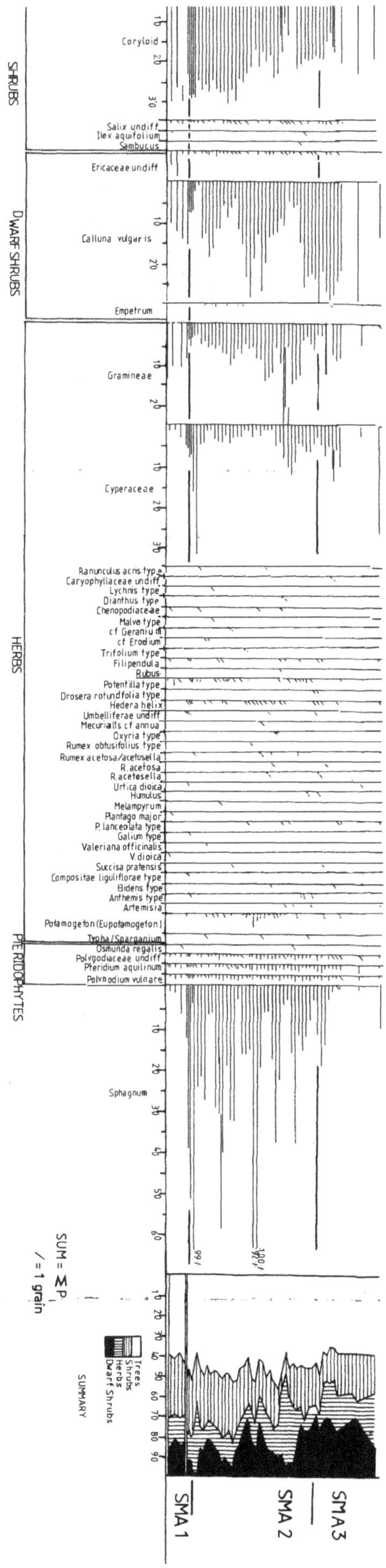

Figure 10 Pollen diagram from Soyland Moor site A
Page 176

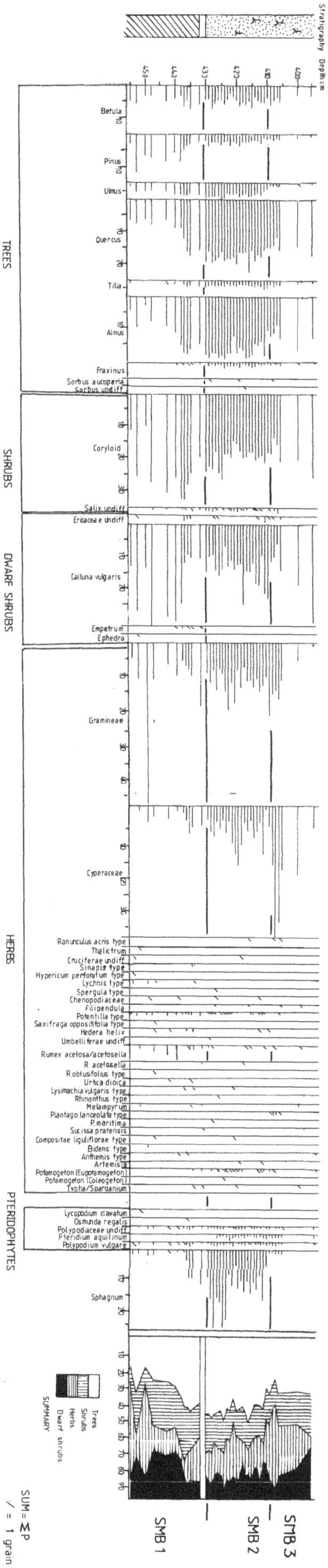

Figure 11 Pollen diagram from Soyland Moor site B
Page 177

Please view pages in original size at this link: https://doi.org/10.30861/9780860543299_original

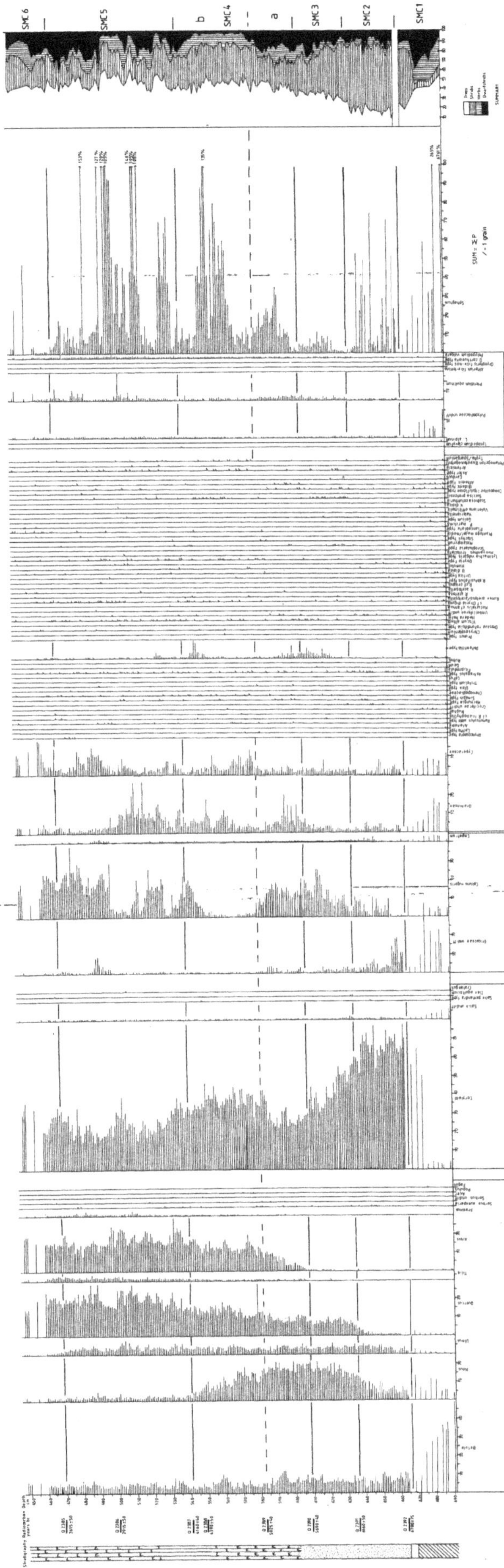

Figure 12 Pollen diagram from Soyland Moor site C
Page 178

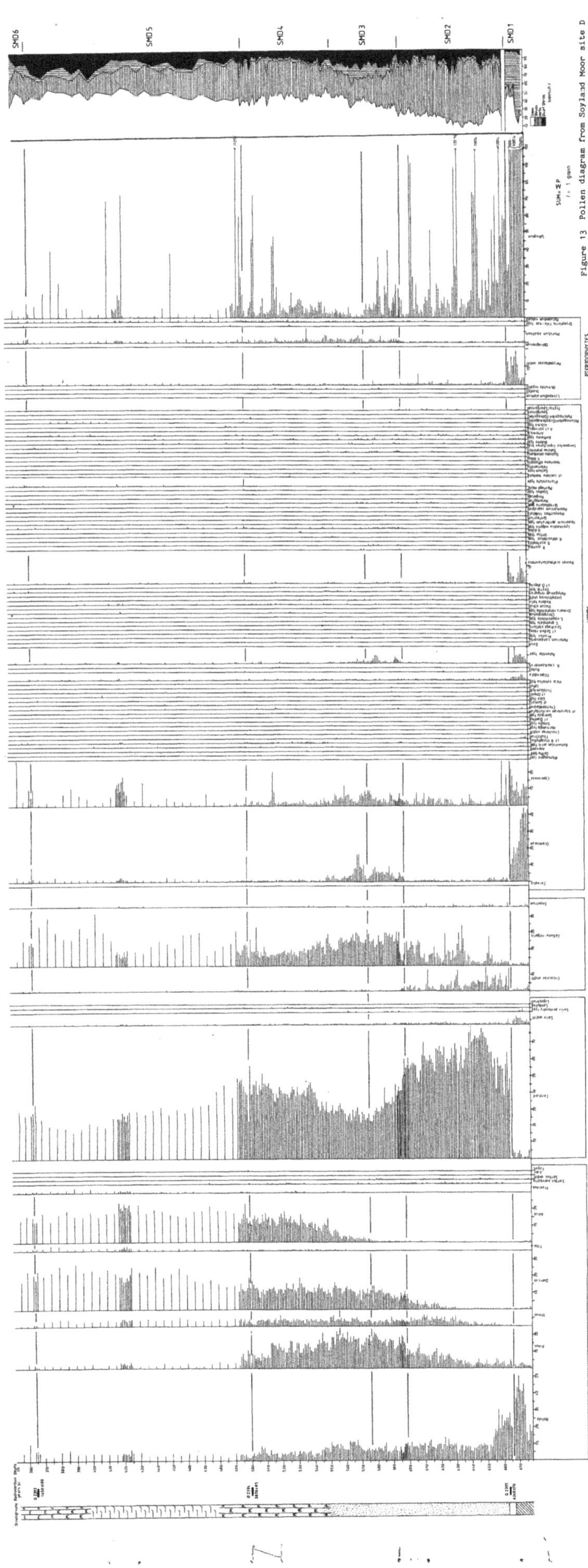

Figure 13 Pollen diagram from Soyland Moor site D
Page 179

www.ingramcontent.com/pod-product-compliance
Lightning Source LLC
LaVergne TN
LVHW071211210926
848482LV00015B/777